T5-ASB-131

A GEOGRAPHY OF
plants and animals

THE BROWN
FOUNDATIONS OF GEOGRAPHY
SERIES

Consulting Editor
ROBERT H. FUSON
University of South Florida

A GEOGRAPHY OF

Agriculture
James R. Anderson, University of Florida
Transportation and Business Logistics
Edwin J. Becht, University of Oklahoma
Plants and Animals
David J. de Laubenfels, Syracuse University
Geography
Robert H. Fuson, University of South Florida
The Atmosphere
John J. Hidore, Indiana University
Population and Settlement
Maurice E. McGaugh, Central Michigan University
Industrial Location
E. Willard Miller, The Pennsylvania State University
Water
Ralph E. Olson, University of Oklahoma
Earth Form
Stuart C. Rothwell, University of South Florida
Minerals
Walter H. Voskuil, University of Nevada

THE BROWN
FOUNDATIONS OF GEOGRAPHY
SERIES

A GEOGRAPHY OF plants and animals

DAVID J. de LAUBENFELS
Syracuse University

WM. C. BROWN COMPANY PUBLISHERS
DUBUQUE, IOWA

THE BROWN
FOUNDATIONS OF GEOGRAPHY
SERIES

Consulting Editor
ROBERT H. FUSON
University of South Florida

ISBN 0–697–05154–4

Library of Congress Catalog Card Number: 73-118882

Printed in the United States of America.

Geography is one of man's oldest sciences, yet it is as new as the Space Age. Knowledge of the earth obtained from satellite photography and measurement, remote sensing of the environment, and by means of other sophisticated techniques are really but a stage in the evolutionary process that began with ancient man's curiosity about his surroundings. Man has always been interested in the earth and the things on it. Today this interest may be channeled through the discipline of geography, which offers one means of organizing a vast amount of physical and cultural information.

The **Brown Foundations of Geography Series** has been created to facilitate the study of physical, cultural, and methodological geography at the college level. The **Series** is a carefully selected group of titles that covers the wide spectrum of basic geography. While the individual titles are self-contained, collectively they comprise a modern synthesis of major geographical principles. The underlying theme of each book is to foster an awareness of geography as an imaginative, evolving science.

Preface

At first glance a study of wild life on the surface of the earth, including the geography of that life, may not seem too important. After all, technological man can control the kinds and distribution of living things, leaving the wild environment as a historical memory of importance only to primitive man. The facts of the matter, however, are that man is far from achieving such control of his environment and his best efforts are showing surprising breakdowns; surprising because there is still a lot to learn about the functioning of life. There are various levels of importance to man of living nature.

The harvest of wild plants and animals continues to dominate the economy in large areas of the earth's surface even if only modest numbers of people are involved. Wild trees yield wood and other products such as quinine and chicle and nuts. Wild animals in a few areas provide an important source of food. Fish, particularly, are taken from nature on the high seas.

For the most part man has applied a measure of control over the natural environment. Forest management is applied to most commercial production of lumber. Thinning, attention to restocking, and erosion control help to insure an optimal and steady yield. Similar practices characterize range management. The manager must know what lives where and in what quantity. The effect of all changes, induced or otherwise, must be understood. Even fishing is being subjected to increasing management. Quotas, seasons, and restocking are beneficial to continued health and yield.

The importance of living things is manifested in indirect ways. There is more to be had from nature involving living things than plant and animal products. Water supply is vital to man and can be greatly affected by changes in the corresponding biological factors. Plant cover retards runoff and animals can change the plant cover. So can man. Recreation is becoming increasingly important as more people come to live in man-made inadequate environments. The role of wild plants and animals in recreation is obvious. Man even manages the plant cover to avoid the erosion that his activities might otherwise induce.

Not enough is known about the reaction of nature to what we do. Our accelerating activities are creating problems faster than solutions are being found. It is easy to understand the menace of erosion. Then there are introduced pests. Is a vigorous plant or animal introduction without its regulators a blessing or a pest? The things which have happened to natural predators is a scandal. They have been killed off purposely or carelessly without a thought to what might result from the loss of their regulating effect. But we find out soon enough. Pollution is upsetting all sorts of balances with results still developing. Not only do we need to know what lives where but we must constantly monitor the changes in order to avoid as much disaster as possible.

All of these comments refer to the larger subject called ecology, the relation of living things to their environment. This book is concerned with the geographic part of the natural environment. Other fields such as biology, agronomy, and anthropology address themselves to other aspects of the subject. The role of differences from place to place in the important ecological problems of nature is detailed in this book, examined at the five levels, population, biota, community, formation, and ecosystem. Essentially the book stands on the ground where geography and ecology overlap and should serve both disciplines.

No book has been written before giving the geographer's view of the biosphere. The many books called variously biogeography or plant and animal geography (phytogeography, zoogeography, etc.) have all been written by biologists. It is hoped, therefore, that a fresh viewpoint has been expressed. There does not appear anywhere previously to have been published a world summary of the floristic realms and phases, what they are and where they are. The analysis of the significant elements of formation stem from original research published at greater length elsewhere. The treatment of sea life and of the ecosystem here is weak because these areas have not been developed as fully by geographers, but there may be an element of poorer knowledge of the author being manifested.

All of the maps in this book were executed by John Fonda, staff cartographer of the geography department at Syracuse University and benefit from his skill in translating a rough compilation into an effective product.

Contents

Chapter **1**

Introduction

There are important differences from place to place in the plant and animal content of the biosphere and it will be the objective of this book to illuminate these differences. The biosphere is that zone on the interface between the atmosphere above and the earth below normally inhabited by life. To be sure, the biosphere also includes the oceans, but it is not feasible at this time to give the same treatment to the geography of the oceans as is possible for the land. Furthermore, it is the land still where most of the activities of man are located, and the ultimate purpose in studying the geography of plants and animals is to understand the habitat of man. More directly, the intent here is to identify the patterns and distributions of plant and animal life, to suggest explanations for these geographical variations, and to consider some of the implications that may be attached thereto. The title, by referring to geography first, emphasizes that patterns and distributions are the focus of attention in contrast to the orientation of biogeography which by tradition has directed attention primarily to the organism. Five subsidiary but interrelated themes have been developed that relate to the geography of plants and animals, beginning with *individual populations* (taxonomic units), followed by *regional assemblages* (floristic and faunistic provinces), then *direct associations* (life communities), and *structural complexes* (formations), leading finally to *life systems* (biosystems and ecosystems).

Historical Background

The study of the geography of life has an interesting history with a remarkable diversification within the last hundred years or less. Earlier very little distributional information was available except at the local level. A further difficulty lay in the lack of any system for classifying living things, a deficiency first overcome by the work of Linnaeus in the latter part of the eighteenth century. Linnaeus was stimulated by the vast accumulations of specimens and observations stemming from the world-wide explorations before and during his time. Subsequently there began to appear maps showing floristic and faunistic regions. By the middle of the nineteenth century some considerable regional detail was possible as in the work of Grisebach. In the meantime Humboldt, the great naturalist, had proposed that the world's plant life be studied in terms of form. Thus a floristic assemblage was seen to be united in terms of similar or complimentary forms. The ideas of Darwin had a profound influence on the study of biogeography (plant and animal geography) because of the link he made between living things and their environment. Life forms developed in response to environment, and the two should vary in harmony from place to place. There followed intensive studies on the physiology of life but the botanists and the zoologists worked largely independently of one another. The monumental works on plant geography, particularly those of Warming, Graebner, Schimper, and Drude appeared at the end of the nineteenth century. The important works of the zoogeographers, from Wallace to Hesse, roughly span the same period. It was these contributions which led to the diversification of the field more or less into separate disciplines.

The most conspicuous new discipline related to biogeography to appear at the end of the nineteenth century is the study called ecology. The all-inclusive theme of ecology is the relation of living organisms to their environment, only one aspect of which is the study of patterns to illuminate relationships. Nevertheless, because ecology is such a vigorous field, its overlap with geography is considerable. Ecologists by inclination work very close to the individual organism and so ecological geography consists mainly of the local spatial aspects of particular life communities. Less attention has been directed to the patterns of life form, although the definition of ecology is broad enough to include any aspect of the geography of life.

The field called biogeography has taken on its present character largely as a result of the subtraction of ecology. There is a further profound division of biogeography into plant geography and animal geography with minimal communication between the two. Modern biogeography is particularly concerned with the patterns of species and of floras and faunas, the explanation of which can never really be divorced from ecology, even though there still remains a great deal of purely exploratory work to be done.

A third approach to the geography of life of particular concern to geographers is the study of vegetation formations, but it does not enjoy the status of a discipline. Ideas about regional vegetation formations have been passed back and forth between geography, climatology, ecology, and even soil studies. The reason for this derives in part from the intriguing concept of natural regions. Particularly in the earlier part of the twentieth century, there was great enthusiasm in all the fields just mentioned for such a unifying concept, to a point where (for some workers) the climate, soil, plant species, and plant forms together were just aspects of a super organism, the climatic climax formation. Allegiance to the natural region concept is waning but its effects on vegetation studies are still strong. Many maps in current use are illogical unless it is realized that they were originally produced to illustrate the vegetational aspect of the natural region. Vegetation formations should be distinguished independently of other factors particularly because a major goal for them is as indicators of other factors.

Interest now centers on whole life systems in space and further on life-environment systems. The latter is the ecosystem, a softer concept than the earlier organismic ideas. Analysis of the nature of ecosystems is still in an early stage and, although it is a useful concept, it is well supplied with pitfalls for the unwary. That there are interactions is not questioned; the problem lies in the nature and strength of the connecting bonds. Only tentative statements can yet be made concerning the geography of ecosystems.

Subdivisions

It has already been stated that there are five themes in the geography of life and each will be the subject of a separate chapter. There is an ascending order of complexity of the content of space among the five themes involving a different sort of approach at each level, always including the common geographical element. For each theme, three steps may be applied, the first of which is largely descriptive. One must begin by identifying the patterns and distributions of the factors under scrutiny. The second step confronts explanation: how did the patterns identified get that way? The final step is to explore the significance or importance which can be attached to the geographic aspects that have been illuminated.

The simplest kind of geography of life deals with the patterns of single breeding populations, the subject of Chapter 2. It is the task of the taxonomist to identify and distinguish the discrete breeding population which the biogeographer can then map; but, conversely, the biogeographer may clarify through maps the population relationships for the taxonomist. It is mainly these specialists who are prepared to construct the geography of populations, but the results are also of interest both to those concerned with changing environments and to the cultural geographer.

The aggregation of all the taxonomic elements within a region, whether rare or common, specialized or tolerant, is called the *biota* (flora when plants are listed and fauna when it is animals). A matter of scale is involved here because individual species may extend over a considerable area, even when not occupying all parts of it. The study of biotic provinces is carried out at rather small scales (large areas) and by the same specialists who study individual populations. Such studies have led to some rather powerful ideas about what has happened in the past on the earth as well as to insights into the way life has diversified in relation to the environment.

When organisms are studied more intimately they are found to be interacting with one another in particular combinations or communities, suggesting an important way to look at living things and their environment. The distributional aspects of life communities (or *biocenoses*) tend to appear at rather large scales (small areas) and deal more with specific environment-life relationships than do regional studies. Thus it is the ecologist who produces most of the information concerning areal associations of living things as one part of his broader pursuit of understanding.

The life content of area can be presented in terms of physiognomy or physical structure which does not necessarily correlate with patterns of biotic associations. Because plants add up to the greatly preponderant bulk of living things and are furthermore fixed in position as individuals, the physiognomic structure equates essentially with the vegetation formation, even though animals have a part in the total picture and contribute in indirect ways to the vegetation form. Individual species have their characteristic form, but nevertheless the important species may vary considerably in form as well as in range through a variety of environments. As a matter of fact, the very plasticity and tolerance displayed helps make such species important. Thus it is that form must be studied separately from taxonomy. The development of the concepts of vegetation formation has been shared by a variety of specialists as has already been seen. The integration through the aggregate of plant structure in an area of a wide range of environmental factors has made vegetation formations a matter of general interest and is the reason why vegetation studies are not strongly identified with any particular discipline.

The most complex interplay of living things in space, the total content of life in an area, can be described as the biosystem or *biome,* the life part of the ecosystem. The biosystem includes all aspects of interrelationships, the dynamic life system of a place. Because it is the most complex way of looking at life it is the least well articulated but probably the most important. The ideas of systems analysis, being applied here, allow something less than organismic structure to be conceived as a system and it is hoped that, as the ecosystem concept develops, it will lead to highly accurate generalizations about the organization of living things in space.

It must be recognized that before any study of distributions can be pursued, the pattern of distribution must first be made available. A great deal of effort is required to assemble the necessary data because even a simple pattern involves many locational facts. Occasionally, as in England, a coordinated effort is made to produce an inventory, but more often the facts are won by piecemeal efforts. Frequently when the range of some population needs to be known, it has to be mapped for the first time, an encumbrance to any study which has this need. Somehow the basic data must be produced and this is the first step in the geography of plants and animals. The next step deals with population patterns taken individually after which combinations and interactions can be considered.

Chapter **2**

Individual Populations

Many thousands of living species inhabit the surface of the earth, each with a distinct pattern of distribution, no two of which are exactly the same. Furthermore, new species are still being discovered, so it follows that the area occupied by each species is not known in every case. Because of incomplete knowledge and because of vast numbers, a consideration of the geography of taxonomic units must deal with examples, of which there is more than an adequate supply. In fact, the available information is enormous and the problem is really only one of organization. There are in the literature many maps showing different distributions which have often been classified into types of pattern. Much can be advanced through theory or experiment to explain both individual patterns and classes of pattern and this in turn leads to important implications and conclusions. Many factors go into permitting or denying area to living things, and the study of biological distributions illuminates both biology and the study of any factor which affects biological things.

Taxonomic Patterns

In dealing with the geography of individual populations the first step is descriptive. The area occupied by a population must be mapped and the essential characteristics of that distribution can then be identified.

methods of representation

Ideally, a map of the exact area occupied by all the members of a population should be made. In dealing with any considerable area, this is not possible and some kind of a generalization must be made. The more important of these can be described.

The most primitive distribution map is one which shows by *unit symbols,* such as dots, the individual locations where the species under consideration are known to exist (See Map 2-6). There may be very few known collection sites but such as are known can block out, more or less, the range of the population. A major drawback of such scatter maps is that they often reveal more about the pattern of collecting, where the collectors have been, than where the species lives. Nevertheless, a scatter map often even suggests internal variations within a range pointing to ecological relationships.

With greater amounts of information, the map maker may be emboldened to outline a *range,* thus suggesting the outer limits of penetration (See maps 2-1, 2-2, 2-4 and 2-5). When a range map is drawn on the basis of scanty information, as sometimes happens, it can be quite misleading. The range map obviates the frequency of collecting factor but also loses all internal distributional information. Occasionally, however, secondary range factors are added, such as areas of dominance, areas of greater frequency, or areas of commercial potential.

For greater accuracy, some kind of *grid* is used. The most convenient grid is some existing areal subdivision such as counties. Information is gathered with respect to each grid unit and the resulting map shows all of the units within which the species is known to live. Geometric grids, as in the *Atlas of British Flora,* have been applied in a few cases. The finer the grid, the more information that can be extracted concerning relations with habitat.

The most detailed map divides the range of a taxonomic unit into the actual *areas occupied* and the areas not penetrated, generalized with respect to the scale used (See Map 2-3). This is important for populations that are quite discontinuous, such as those with strong habitat preferences. The degree of generalization used in mapping by area occupied, or by any other method used, can vary broadly and cause much overlap between the information displayed by one method compared to that of another. It is important to appreciate the limitations inherent in the kind of distribution map being used.

classes of pattern

Many types of pattern can be singled out among the thousands of examples that have been constructed and quite a variety have been described. The more categories, the more detail can be accommodated, but above all what is needed is a limited set of important classes, of which five are described below. These, then, can be further subdivided as needed.

Cosmopolitan

Those plants and animals found in most parts of the world are referred to as cosmopolitan species. There are several types of these ubiquitous species. On land, certain plants whose propagules are as dust and whose habitat tolerances are widely satisfied include cattails and other marsh plants, some ferns and other primitive organisms. The sea offers easier opportunities to many simple organisms. More recently man and his followers such as flies, mice, and the like, plus such weeds as plantains, dandelions, and sour clover have penetrated the far corners of the globe. If consideration is raised to higher taxonomic levels, the attribute of cosmopolitanism can be applied to many things.

Continuous

Many distributions fail to become worldwide but nevertheless extend continuously over a broad range. Such widespread species can be grouped into a major class. There are two prominent subtypes of continuous distribution. In some cases a population may be continuous throughout the major part of a continent, while in others the range may be zonal as throughout the tropics. In the latter case, terrestial ranges are necessarily divided by ocean gaps within their zone. Again, more examples of genera and higher groupings are available for continuous zonal ranges than are individual species but also again the actions of man have provided many new examples. Innumerable species are found over the greater part of a continent such as the mountain lion or the box elder. Zonal elements include the moose (Map 2-1) and the common juniper at higher latitudes or certain mangrove species in the tropics.

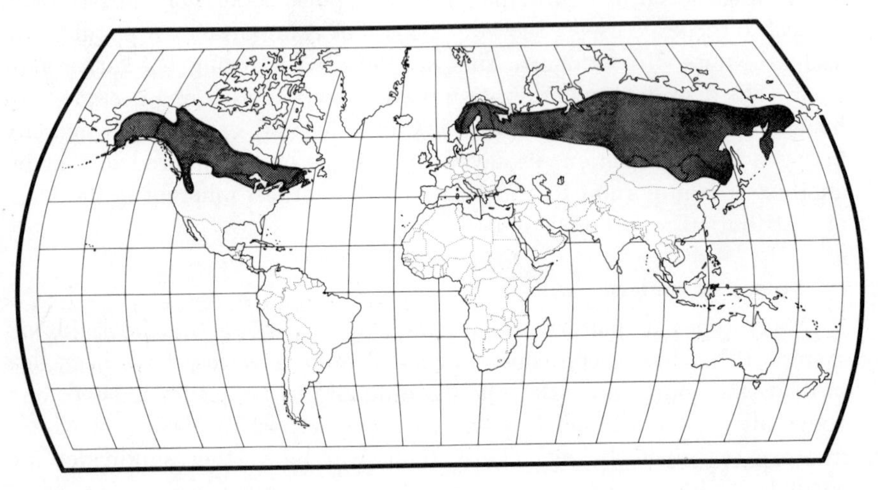

MAP 2.1. Range of moose (**Alces alces**), an example of a continuous zonal range.

Discontinuous

It is quite usual for a population, although occupying considerable area, to have one or more gaps between occurrences. Gaps can represent incomplete knowledge, but real gaps are common. The lion is an interesting example of an animal whose range is discontinuous, occurring in Africa and India, but the gap may be the recent handiwork of man. The sweet gum has a discontinuous range in southeastern United States and in Central America (Map 2-2). Many lesser examples, of course, exist.

Scattered

When a species is only locally represented at best but does occur in a number of places, it can be said to be scattered (Map 2-3). This sort of pattern differs only by degree from discontinuous and refers primarily to organisms with rather special habitat requirements. An example might be the cactus of southeastern United States.

Endemic

The most restricted class of patterns is called endemic and refers to a population at home in a single small area or cluster of areas. How small an area depends on the point of view so that on a worldwide basis endemic might refer to Australia, but usually a much more restricted area is meant. Islands are famous for their endemic plants and animals, particularly the more isolated islands. Endemics occur on the continents as well, which raises many interesting questions. The catalpa in America is an important tree whose natural range is quite limited (Map 2-2) and one would have to consider the whooping crane in this category even though its range was larger not so long ago.

Other kinds of pattern tend immediately to suggest certain correlations. An example of this is a seacoast location. Actually the geographer is most interested in correlations, and any areal relationships which can be deduced point toward explanations of pattern and the dynamic interaction of the organism with its environment.

Factors in Distribution

Explanation of a simple biotic distribution pattern can be highly complex and can defy complete rationalization. There are many factors which are relevant to why an organism occurs in one area and is absent in another. These can be organized under seven headings.

climate

Manifestly, climate is the major factor for the general limits of most living things. Taken outside of their general climatic milieu they can

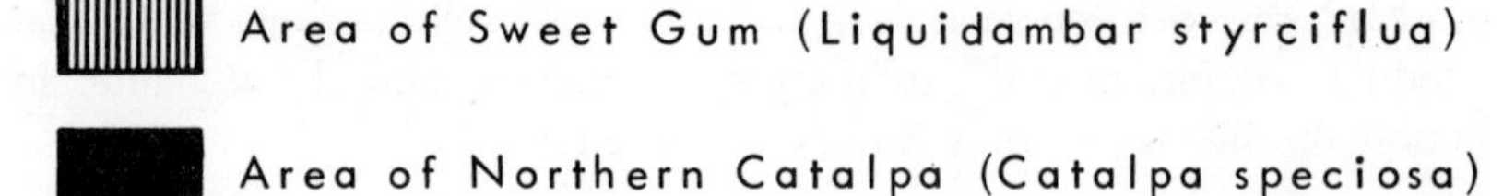

MAP 2.2. Distribution of sweet gum and of northern catalpa, showing a discontinuous distribution and an endemic (localized) distribution.

not endure. Just how organisms relate to the elements of climate, however, is not easy to establish, partly because the importance of climate may vary with the life cycle or between individuals. The relationships, furthermore, are often complex and are, of course, affected by nonclimatic factors. With important short-lived species, experiments in a

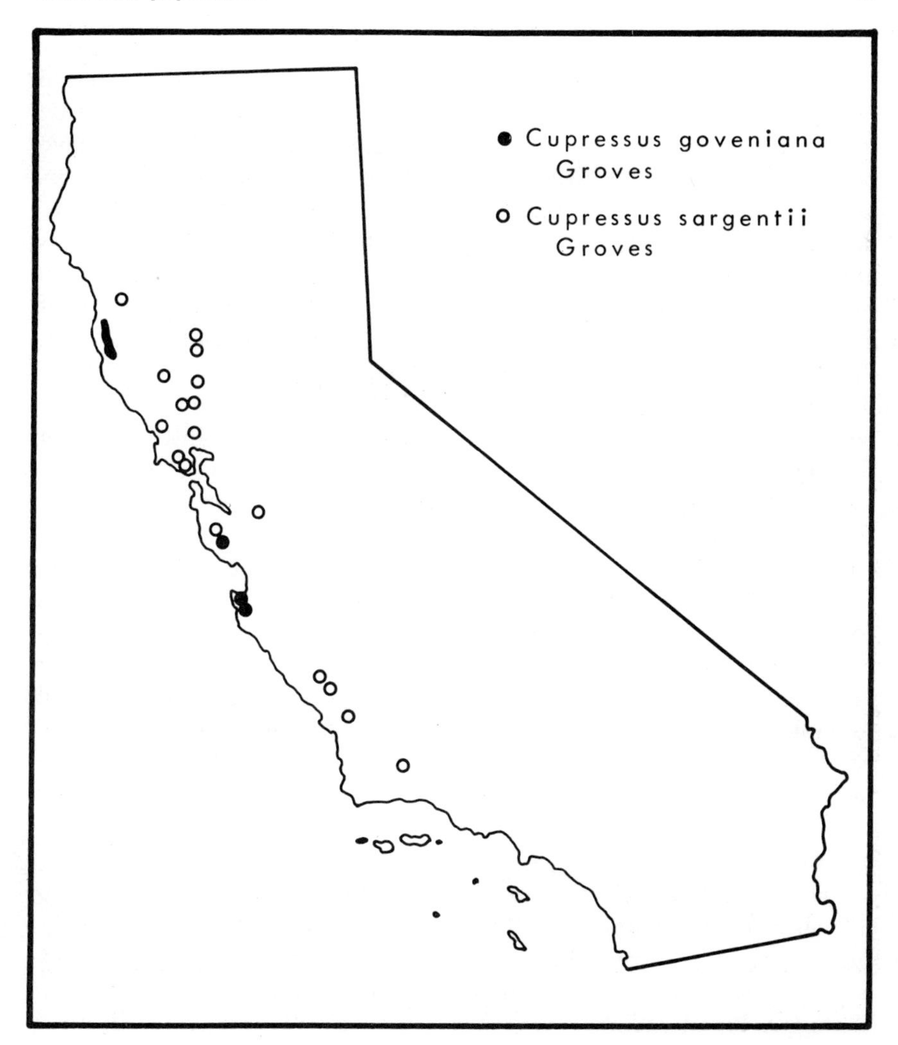

MAP 2.3. Distribution of two California cypresses showing scattered distributions.

laboratory can be made; large long-lived species are generally impractical for use in experiments. Much is known about the geography of each climatic factor, and where climate is important one can turn immediately to climatic maps to discover the character of the distribution in question.

Temperature

In one way or another, temperature affects every organism. Some species are killed by frost; others have come to depend on frost in their life cycle. There is a great variety of critical temperatures which either must occur or must not occur in order that some part of the life cycle or even life itself may be sustained. Orange blossoms are eliminated by the least frost but cherry blossoms will only develop if the buds have been adequately chilled. Elephantiasis is a tropical disease which disappears where cold is encountered. The critical effect of temperature may be the accumulation of degrees or the duration of temperatures above or below some particular amount. This is best known for crop plants where, for example, grapes or cotton must experience a total of a certain number of minimum-degree days in order to ripen their seeds. For frost-kill it is often not the lowest temperature experienced but the duration of freezing that is operative. Not only are the critical values not always known, but also information concerning the distribution of the corresponding data is seldom available. Therefore it is neccessary to use the kinds of crude data that are available, and for the most part this means average temperature. Thus the patterns of distribution are compared with the patterns of isotherms to suggest in a general way what the temperature limits for different species are. The correlations so obtained are poor enough that their is frequently doubt that any meaningful relationship has been suggested. Nevertheless, many rather distinct temperature-related limits can be demonstrated. A widely used example of this is for the sugar maple whose range lies roughly between the isotherms of minus ten and minus forty degrees Centigrade for mean annual minimum temperature (Map 2-4).

Moisture

The importance of moisture to plants and animals is of equal significance to temperature, but this means moisture effectiveness and not brute amount. Moisture effectiveness is based upon need which is not a simple value to derive. Need is essentially a function of energy, including a broad radiation spectrum together with heat, plus diffusion rates and probably other factors. Science is far from an exact expression of these parameters so that need must be based on difficult empirical measurements or calculated approximations based primarily on temperature. Animals, being mobile, can seek out necessary water supplies, but in the food chain they are ultimately dependent on plants and do not escape the effects of moisture shortages. Given the running values of need, shortages can be absent, intermittent, or continuous. All organisms have some limiting moisture stress beyond which they must either become inactive or die. A camel can endure longer periods without water than most animals and a maple tree could wilt next to a flourishing mesquite. Complex moisture requirements, similar to temperature re-

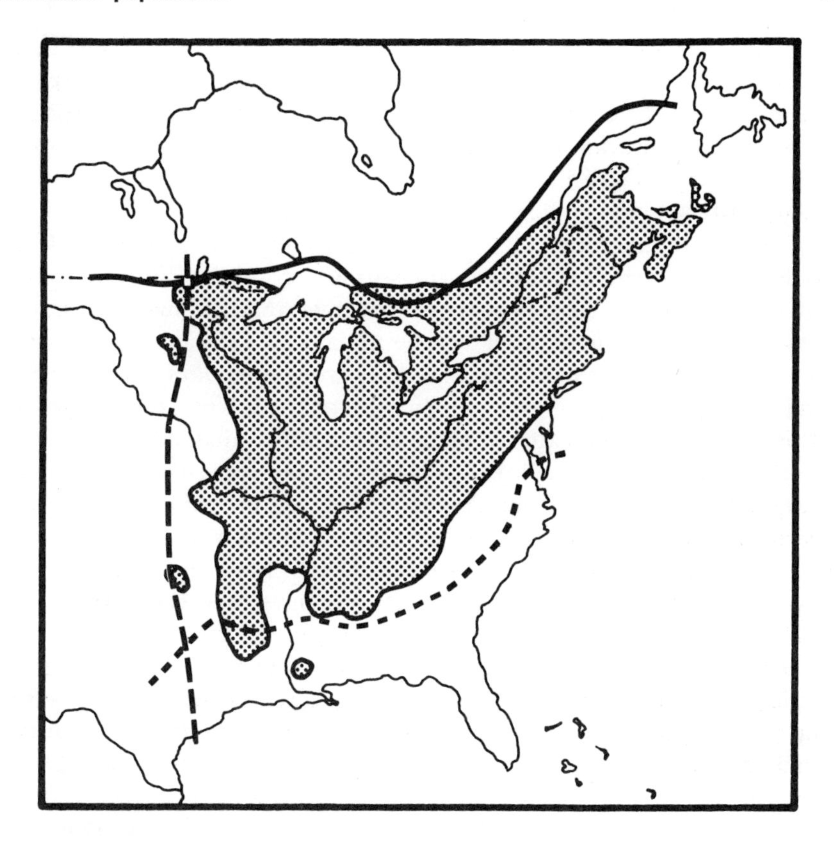

Moisture Index = 0 (Thornthwaite)

Average Annual Minimum Temperature = -10°C

Average Annual Minimum Temperature = -40°C

MAP 2.4. Relation between the distribution of sugar maple and several parameters of climate. The lack of a closer fit may be explained in part by local edaphic conditions.

quirements, also exist. For example, many desert ephemerals must have their seeds soaked a certain amount before they will germinate. Even with crude moisture zones a relationship can be seen between species limits and moisture levels.

Photoperiod

The role of photoperiod is that of a key for plants or animals to regulate their cycle with respect to seasons of stress. When a predetermined day length has been reached either through increase or decrease, a new part of the cycle can be initiated. As the days grow short many animals prepare to hibernate and when the days grow long a dandelion sets flower. It is not always apparent what the triggering factor is, because temperatures or in some cases available moisture supplies fluctuate in phase with the photoperiod, but experiment has shown in many cases that day length is the key. A case in point is that many greenhouse plants seem to know when it is spring without being subject to outside air conditions. Organisms keyed into photoperiod, unless they develop variants with different responses (ecotypes), are restricted to particular latitude zones.

Directly or indirectly, other climatic elements have a bearing on distributions. Wind has a drying effect and, where strong and persistent, a mechanical effect which could be inimical to some organisms. Humidity also makes a difference. There are places where desert plants grow on the basis of dew alone and little animals lick up the dew to satisfy their water needs. Fog also can be harvested for its moisture. There is no attempt being made here to be exhaustive concerning the geographical effects of climate but simply to suggest the leading factors.

soil

The effect of soil can be very pronounced over a short distance, and often soil factors can recur in a mosaic and thus loom out as of singular importance. The more extreme soil conditions are rarely widespread so that there is a tendency to think of soil as a local factor. Occasionally, as in active areas of deposition, there may be a really large unit. Differences from the general "zonal" soils involve both drainage and chemical content. Poor drainage or excessive drainage, either through lay of the land or through internal soil structure, tend to duplicate drier or wetter climates. Galeria forests along streams extend wetter habitats into drier environments while desert conditions may be seen on dry soils where the climate generally supports a more demanding biota. Deficiencies or excesses of particular chemicals beyond the toxic level are more unique and have important local effects on native plants and animals. Livestock could not be raised in a certain region of New Zealand because of a soil chemical deficiency (since rectified), and salt marshes contrast sharply with other adjacent habitats even those just as wet. It is of interest that in the more favorable climatic zones plants are more sensitive to minor edaphic (soil) differences. For the most part, soil is responsible for local details in the pattern of distribution whereas climate produces the broad patterns.

surface configuration

Terrain has many variations which effect the distribution of habitat factors. In the simple case of plains, terrain is neutral, while rugged landform can make a great deal of difference. The division of the surface between land and water is a part of surface configuration and other effects are analogous. Terrain works in part to create complexities in the climatic pattern or can be associated with soil variations. These complexities can be considered islands, straits, and peninsulas or outliers, barriers, and corridors. Gaps or barriers operate in terms of the effective range of migrations. For islands, another factor operates. A specific population, whether insects, mice, or trees, has a minimum range area necessary for continued propagation. In a smaller area the chances for successful reproduction or survival to maturity fall below unity. Involved are food supply and competition as well as predation. Surface configuration can have a great deal to do with details of species distributions and with preventing migration to distant regions.

disturbance

Disturbance is taken to mean any irregularly occurring event that may affect the growth of an organism including events whose frequency of occurrence can be expressed. Many kinds of disturbance exist including windstorm, flood, drought, fire, disease, grazing, and man. There is a fine line between sporadic drought and seasonal drought, for example, or between endemic parasites and infestations. Thus it is the activities of man, setting fires, upsetting the food chain (as in killing off the predators), or directly attacking the biota, that are most widely accepted as qualifying as disturbance. In any case, disturbance upsets the balance and, for some organisms, opens up niches (a way of making a living), while for others, it may eliminate some critical habitat factor or even kill the organism directly. Vast numbers of opportunistic species depend on disturbance to allow their presence. In the South, controlled burning is used to promote quail population where hunting is desired. Without abandoned agricultural land, white pine requires a fallen tree to throw up the necessary mineral soil for a new seedling to succeed. Some pines require fire to open their cones. Manifestly, disturbance kills off other species. The conservation literature is full of accounts of endangered animals. In the larger sense there is a continual dynamic readjustment of the biota and the soil following disturbance until some kind of equilibrium is established. This change is called succession, each stage of which has its characteristic species. Where a species is found, then, can be profoundly influenced by disturbance.

dispersal

New areas cannot be occupied without some method of dispersal and dispersal takes time. There are two parts to dispersal; the first is travel and the second is establishment. Plants are passive and can be adapted to mobilization by wind or water currents while animals are active and may be able to walk, fly, or swim. Either one may be carried, as by rafts or by migrating animals, sometimes for astonishing distances as can be seen by the arrivals on isolated ocean islands. Once arrived alive, establishment of a life form is not automatic, for conditions may not be suitable for survival. A maple seed blown into a dry region will not produce a maple tree there any more than an orange seed can result in an orange tree in Canada. The more adaptable species including most pests have an advantage where new environments are encountered. Another problem is competition. Furthermore, the new invader must constitute a breeding population. No matter how fat a lone male rat might get, his kind dies with him. Even a pair is insufficient for some species (see critical area above). Many more species could survive in any area than are found there. When the area is beyond the dispersal range of a potential inhabitant, that species can be called unavailable. No matter how successful a mesquite might be in an Oklahoma grassland, if no mesquite seed reaches the site, there will be no mesquite trees there. Nor were there rabbits in Australia before man brought them. And when an island emerges from the sea or a field is abandoned, even those species within range do not immediately occupy the available space. Only through time does a species expand to occupy the potential area available to it. Disturbance and changing environments constantly open new areas, and in this way there is a dynamic factor in biota. Perhaps the most spectacular advances are those of pests like the Dutch elm disease which each year reaches new trees. Most plants and animals hardly occupy all parts of the world in which they would be able to survive. Their range, rather, is restricted by their dispersal ability where static or by time where a species is spreading.

competition

Even where conditions are suitable and a species is available, it may be excluded by competition. The fact that many California cypresses grow on patches of very bad soils must not be taken to mean that these soils are best for them. Effective competition for cypress is lacking on the bad soils but it cannot compete on better soils. When occasionally a cypress does become established in the better soils it grows much more vigorously. The range of a species is seldom the same as or completely included within the range of any of its competitors, particularly where that competition is absolute. Thus a species often survives in only a part of its permissable habitat.

change

It is often said that change is the order of things. Environments change and populations change; either way the relationship between biota and environment is affected including area occupied. For example, two very different kinds of conditions are associated with restricted population areas. A newly evolved species can have a restricted range for lack of time to have reached other areas. New species are generally vigorous and expanding and can be identified by the proximation of closely related (parent) species. An old species losing out in the race to adjust to changing environments, declining in area and abundance, is the other extreme. Old species usually are far removed from their close relatives. Changing environments have two effects on distribution. When conditions deteriorate, species persist as relicts in refuge areas as, for example, isolated stands of jack pine along the southern part of its range. Where conditions ameliorate, species expand, as jack pine along the northern part of the range (Map 2-5). These dynamic adjustments are in part due to changes in the balance of competition and not just in the ability to survive. Environment includes, as well as inanimate factors such as climate and soil, predators and parasites which can also change. Where a species lives, therefore, is explained in part by changes both in the environment, if that has occurred, or in the population itself, if that has occurred.

Any one or a combination of the above influences may play a role in determining the limits of the range of any given species. Probably other factors could be distinguished. Complexity is therefore indicated and it is no wonder that a precise explanation of species areas is not given. On the other hand, with so many factors related to where things live, a knowledge of species distributions should provide insights into many problems.

Significance of Species Distributions

A variety of inferences have been made from the patterns of distribution of individual species. These patterns are full of meaning, first

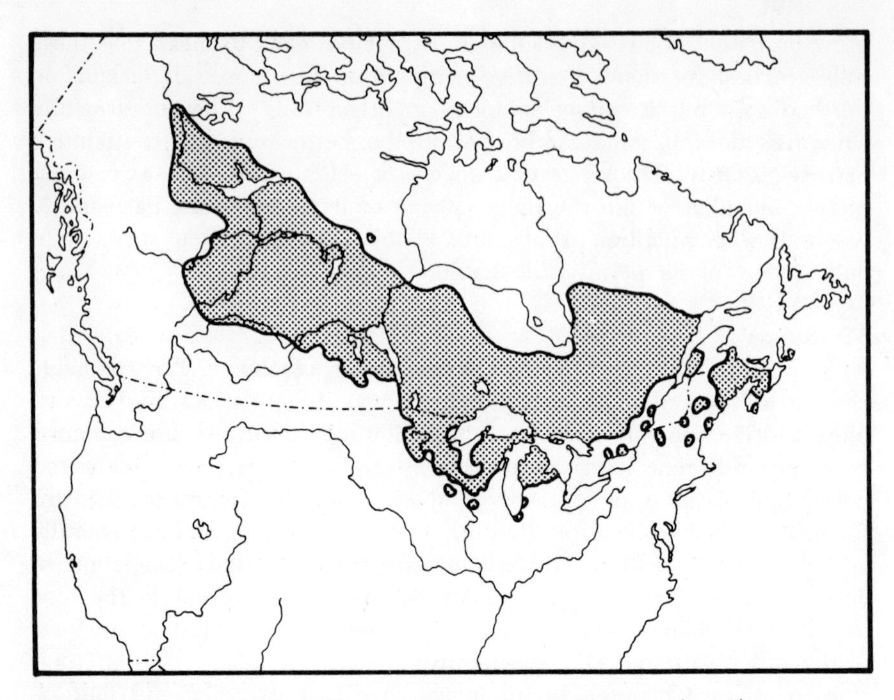

MAP 2.5. Distribution of jack pine showing a fragmented southern area, particularly in the east, and a simple northern boundary.

for the organism itself, further among related taxa, followed by clues to the dynamics of the environment, leading finally to insights into interaction with the affairs of man.

nature of individual taxa

Where a population lives is conditioned by its own physiological characteristics, the other side of the coin from the above discussion explaining population limits. If one can explain the distribution of a population there probably has been reference to response to the environment. The approach to studying physiology through geography can be either from discovering areal correlations, given the population pattern, or from testing suspected relationships by areal analysis. In either case, where spatial relationships exist there is empirical evidence which can be quite powerful. A few examples can demonstrate the range of possible results.

The point of origin may be a map. If one were to compile a careful and detailed map of deer population it would be far from uniform. Two related factors ought to emerge from a study of this distribution. Deer are largely absent in open country and deer tend to avoid unbroken forest. This suggests that deer are edge animals (the facts about deer have, of course, long been known). The distinct tree, *Darycarpus vieillardii,* grows in rainforests on the island of New Caledonia but not in all rainforest areas. By superimposing a map of the serpentine formations over a map of the distribution of this tree, it becomes apparent that it is confined to the serpentine zone (Map 2-6). *D. vicillardii* is a serpentine endemic.

It may be desirable to seek a geographic correlation for a suspected relationship. For example, it is thought that the tsetse fly breeds in brush country. The distribution of tsetse fly and brush can be carefully mapped and compared to see if a correlation results. If it does there would be strong empirical confirmation of the initial assumption. If one assumes that elms grow naturally in freshly disturbed areas, a correlation test can be made. The distribution of young elms, however, correlates with a cover of weeds as along a mowed right of way or on the edge of a field, corresponding to the second or third year after baring the ground surface. The young elm seedling needs some protection.

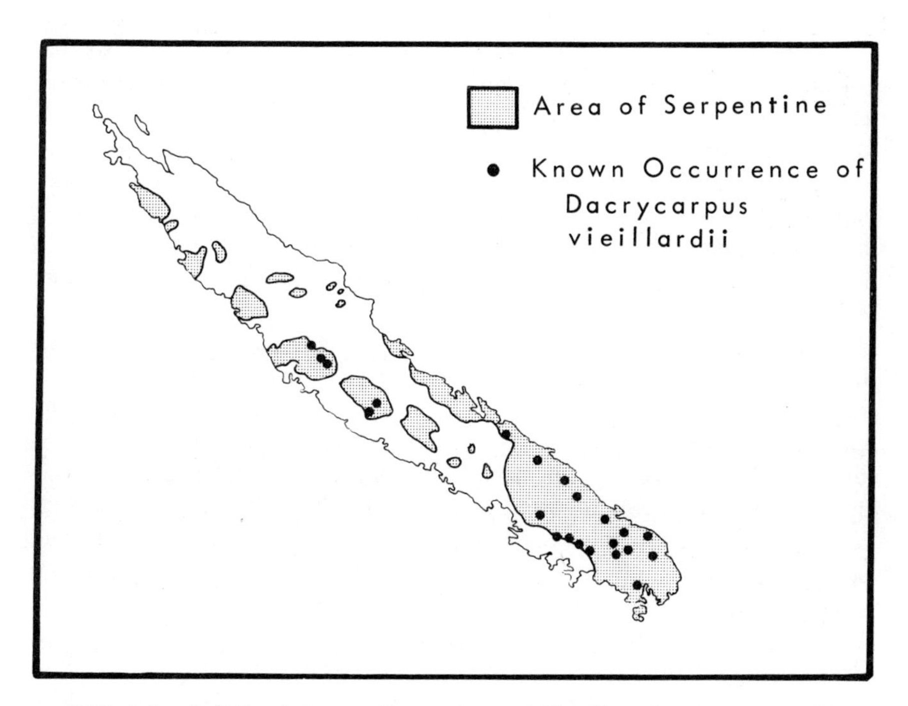

MAP 2.6. Relation between **Darcycarpus vieillardii** and serpentine soils (New Caledonia). This species is not known elsewhere.

What is involved here is the application of geographic techniques to the study of individual populations. Any factor whatsoever which partakes of areal variation can be studied in this way; the above examples merely suggest how such a problem can be attacked. Geographers lament that ecologists only infrequently take advantage of their geographic opportunities.

evolutionary insights

For the biologist, the area occupied by a population can be made to yield information about its evolution. Spatial relationships correspond in part to taxonomic relationships. This may be approached in several ways.

Relationships Between Individual Species

New species evolve from pre-existing species and a major factor in speciation is isolation. Besides special isolating mechanisms such as differing times for pollenation or mechanical barriers, isolation includes areal discontinuity and habitat differentiation (or both). The latter two are expressed geographically.

It is easy to understand how the areal isolation of two parts of a population can be followed in time by divergent genetic change, so that disjunct populations always suggest the possibility of taxonomic differences. Two related species in similar habitats are referred to as vicarious species, and a series of such species (for example, the sycamore in eastern United States, the southern Rocky Mountains, and in California) have been called a phylad (Map 2-7). Blue jays and squirrels are among numerous examples of the same relationship. Vicarious species tend to have a common ancestor which may be one of the species in question.

Genetic change can lead to differential habitat response allowing a population area to be divided between two species at least one of which has become more specialized or a new species may extend the former range of the ancestral population. Darwin's famous observations of the Galápagos finches is an example of habitat speciation or adaptive radiation. Habitat differences can usually be mapped, so that if two putative species can be shown to occupy distinct habitats, this is a supporting argument leading to their taxonomic segregation. A series of white oak species can show both of the kinds of speciation described here. White oak itself is an upland forest tree in the eastern part of North America. Post oak has a similar general range but is more specifically found on drier sites and open fields. Overcup oak (swamp white oak) is completely immersed in the range of these other two species but grows in moist, poorly drained soils. On the other hand, Gambel's oak grows in the mountain states, valley white oak in California, and Garry's oak in the Northwest. When growing in the same region, closely related species often occupy adjacent habitats. In fact if there is anything to the concept of a "natural genus" it is that a group of related species have differentiated with respect to a variety of ecological opportunities.

Hybridization

Natural hybridization including polyploidy occurs frequently and complicates the problem of classifying living things. The biologist both seeks clues suggesting the hybrid origin of any taxon (taxonomic element such as a species or a variety) and, given a hybrid, he seeks the parents that produce it. Geographically the evidence for hybridization is the appearance of a taxon only in the areas of overlap of two other related populations. For example, *Quercus undulata* is found where the deeply lobe-leaved Gambel's oak (*Q. gambeli*) shares its range with one of several nearly entire-leaved related oaks. This variable and problematical species has been shown to be a series of hybrids and not a valid species in itself. Hybridization is also important in gene flow or exchange where two populations are in contact. This can be seen in human populations where characteristics like round-headedness and type B blood are diffusing into Europe from Asia. Intergrading populations could just as well be differentiating as converging, but in either case the pertinent patterns illuminate the relationship.

Center of Origin

The biogeographic literature is studded with references to the center of origin for groups of species. The value of identifying such a center is that earlier relationships in the early center can be traced, leading to further understanding of development. It is fairly easy to discover the center of variation for any taxonomic group but this is not the same thing as the center of origin. Speciation (adaptive radiation) may progress in an area of variety while older or ancestral species may become extinct elsewhere. For example, a group of needle-leaved *Dacrydium* species form an important element of the Malesian flora, being particularly common in Borneo and New Guinea, and yet their origins, as the needle leaf would suggest, lie in austral zones where today only one species exists. There is a different group of the same genus as well as other related more archaic and endemic genera in the cooler areas today, but the moist, tropical, mountainous zone offers a rich choice of habitat for which many species have differentiated. A center of variation *can* be a center of origin but this is not necessarily true and it is all too easy to make unwarranted assumptions by correlating the two. Evidence for center of origin comprises such patterns as converging

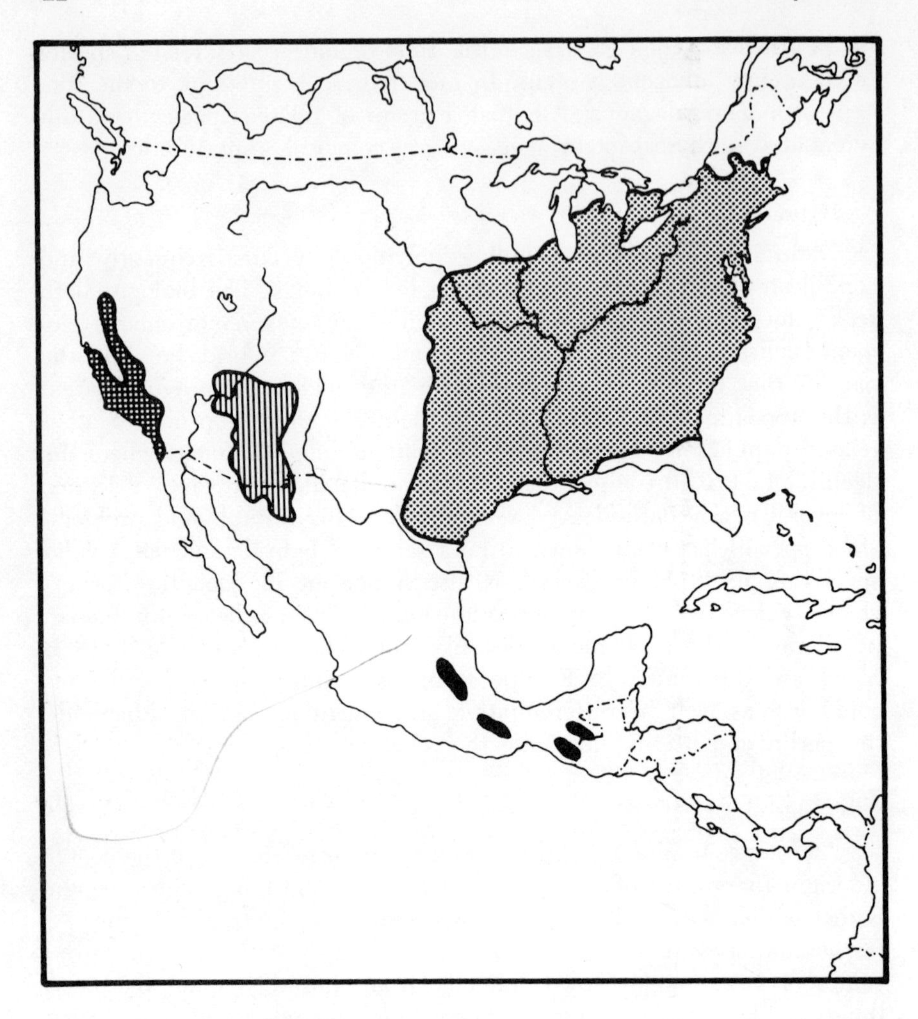

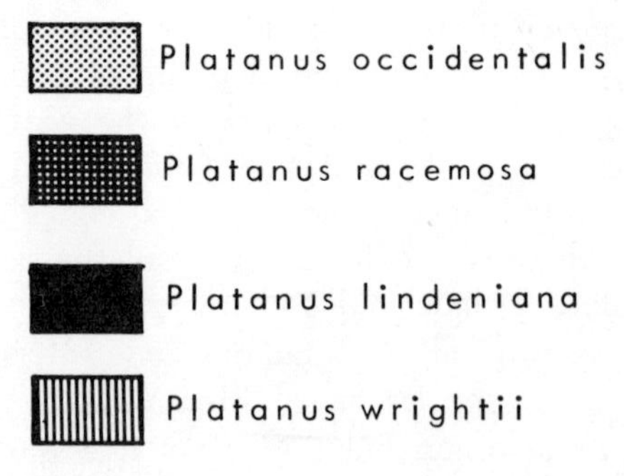

MAP 2.7. Distribution of a set of vicarious species of sycamore in North America.

lines of variation and parallel dispersal of other taxa. The location of relict ancestral forms is also suggestive.

changing environments

Distribution of any population is related to environment so that if environments change there will be inevitable responses in the range of the population. The resulting patterns can be read for a variety of clues that suggest what has happened.

Changing Climate

Change in climate can be thought of, at least in part, as a shift in position. Zones of temperature or moisture belts can and have been displaced particularly with respect to the glacial periods of the recent geologic past. The effect on living things is to cause them to migrate (or disappear). There is a marked geographic expression both before and behind a moving population region. With deteriorating climate, living things become isolated in refugia such as on mountain peaks or in moist basins. Mule deer in the mountains of Southern California are more or less relicts as are monkeys at Gibraltar. The forests in general with their associated flora and fauna in the western United States represent remnants of formerly more widespread stands. The advancing edge of a population is, conversely, not fragmented as there has been no opportunity to populate outlying favorable areas. Therefore, where a population distribution is fragmented with an irregular boundary and with isolated outliers, an unfavorable change in climate may be suspected; an advancing front needs to be documented by evidence of invasion. Because climate is significant to many other things such as geomorphic processes and primitive culture, evidence for past climates and change is in great demand. In fact there are vast numbers of cases where distributions suggest changes in climate, a circumstance to be considered in the aggregate under biota.

Changing Land Area

The most obvious change in land area involves the loss of former connecting corridors or the creation of new links, either of which can be reflected in species distribution. The presence of tree kangaroos and cassowaries in both New Guinea and Australia is evidence of a former connection because neither of these animals is capable of migrating across significant expanses of water. Neither has reached Celebes. The continued expansion of the armadillo into southeastern United States reflects a geologically recent connection between North and South America which was formerly missing. Even more long-range former connections are suggested by closely related species as the American and African mahoganies. On the other hand, the lack of similarities on either side of a gap indicates that the barrier may have been of long duration. Such is the case with respect to Wallace's line (See Map 3-13) where, on both sides, many species' ranges terminate. Differences in land area are important to evolutionary processes as well as being of significance to an understanding of geologic change.

Disturbance and Competition

Two other environmental factors that may change and therefore affect the distribution of a population are disturbance and competition. The results are apt to appear much the same as the effects of changing climate. Disturbance, particularly by man, has condemned many plants and animals to refugia if they are to survive at all. Large animals like the rhinoceros hold out only in isolated territories while in agricultural areas many trees cling to farmers' woodlots. Competition is most noticeable in remote islands where a limited flora and fauna can be easily challenged by robust imports. The Tasmanian wolf may be in the process of being displaced by dogs; and eucalyptus is taking over many forest areas in Hawaii. The effect of imported honeysuckle in the American South on many small plants is profound. Many relict species may have been defeated by a combination of competition and changing climate. It is important to remember that a pattern of and in itself does not give a final answer as to what has taken place but it does supply strong circumstantial evidence to be coupled with other information leading to a firm conclusion.

cultural interaction

Wild plants and animals are significant to the activities of man. Roughly, one can speak of the future value to man, the results of man's past practices and evidences of cultural origins.

Man uses both wild and domestic plants and animals. The wild environment has useful as well as troublesome species whose distribution determines the potential of an area. Foresters particularly have an interest in the inventory of commercial tree species, the object of a timber cruise. A negative effect is produced by pests, a spectacular example of which is the tsetse fly. Primitive man was vitally concerned with where his food supply was to be found whereas modern man's interest may be related to recreation. Conditions and needs change and many areas are still imperfectly known so that the assessment of area for its potential is of great importance. Early man knew how to evaluate an area by use of indicator species and the technique is still valuable. Mullen is a sign of an over-grazed field and the eastern red cedar has a strong affinity for limestone soils. Even geologic exploration can make use of plant patterns.

Man has left a trail behind in plants and animals which can be revealing of the past. Introductions can make over an entire region. Many a "goat island" has had uncontrolled goats devastate its flora. Deer in New Zealand and New Caledonia have had important effects as a local resource. More intimate are the camp followers, bugs and weeds and other nuisances. These are concentrated around or even limited to settlement areas. Certain weeds and shrubs may persist long after a settlement has been abandoned, shouting to the knowledgeable that here man used to live. Lilac and red cedar play such a role in eastern United States. Trees whose products are useful, furthermore, may be left during agricultural clearing. Apparently the Mayan farmers preserved the sapodilla tree known for its fruit and as a source of chicle. The farmlands long ago abandoned by the Mayans can still be traced by the abundance of sapodilla trees in the forest. On the other hand, useful timber trees are selectively removed from forest areas and some experts think that the extinction of large animals in the fossil record can be related to the advent of efficient hunting cultures. In one way or another, a careful study of species distributions can reveal a great deal about the cultural past.

Because the domestication of plants and animals was such a significant milestone in the cultural evolution of society, it is of great interest to locate the hearths of domestication. Both the origins of domestication itself and secondary additions associated with particular groups of men are of interest. The process begins by identifying, where possible, the wild ancestors of the domesticates and then mapping their areas of spontaneous occurrence. If several are included in a complex, this leads to the search for areas of overlaps in which places it is probable that the conversion of wild species first took place. For example, in southwestern Asia, early neolithic cultures had domesticated wheat, barley, lentils, cattle, sheep, and goats, among others. The particular wild strains involved here overlap generally in an area from eastern Anatolia eastward into Iran. It is probable then that the mesolithic people of that general region were the first to gain control over their food supply and all that it means to a primitive economy. The idea and the domesticated plants and animals themselves subsequently were carried far and wide, well beyond the natural range of the wild relatives from which the stock derived. The knowledge of origins contributes to self understanding; species distributions can be used in illuminating origins along with the other uses which have been detailed here previously.

Chapter **3**

Biotic Patterns

Species content of an area is called its biota, flora when plants are referred to and fauna when dealing with animals. It is found that groups of species are often broadly associated with one another over large areas, or closely related species may alternate so that the local biotic inventories of large regions contain most of the same or corresponding species from one to another. Other regions may include few or no relatives. The regions of biotic interrelationship have been given a great deal of attention because of the evidence they yield about past development and interaction. There are similarities between terrestrial floristic and faunistic distributions but there are also great differences due to relative mobility and rates of change so that terrestrial flora and fauna are best described separately after which their origins and the implications of their patterns can be considered jointly. Marine biota have an essentially separate existence but do partake of many similar characteristics. Cosmopolitan elements will tend to be overlooked in developing the regionalization of biota.

Flora

The world can be divided into several major floristic realms (kingdoms or regions), each of which can be subdivided into two or three phases or subrealms which then can be divided into provinces. There never has been complete agreement on the way these divisions are to be made, partly because of various degrees of intergradation, but the important facts of distribution are firmly established. Several dozen floristic provinces can be distinguished such as the Amazonian or the Pacific, Atlantic, and interior North American provinces. The emphasis here will be on the larger divisions, the realms and the subrealms, of which four of the former and ten of the latter are distinguished. The four floristic realms are the *holarctic,* largely in the extra-tropical part of the northern hemisphere; the *neotropics,* which largely coincides with Latin America; the *palaeotropics,* here restricted to Africa and Arabia; and the *austromalesian,* extending from India to New Zealand with an outlier in southern South America. The subrealms are related to moisture or temperature poles within each realm such as the arid and the humid phases of the neotropical flora (Map 3-1).

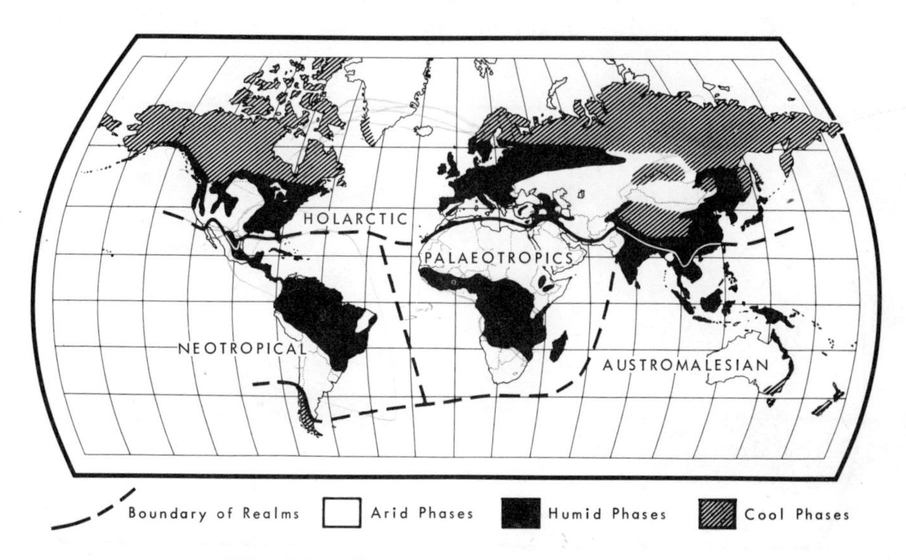

MAP 3.1. Floristic realms and their phases.

holarctic

The holarctic region derives its name from its extension across the nearly continuous land areas surrounding the arctic zone even though in latitude this flora ranges across the Tropic of Cancer in the highlands of Mexico and southeast Asia. The richest variety of species within the holarctic is found within the warmer, wetter parts as might be expected, diminishing and becoming modified both toward the drier and toward the cooler extremes. Thus, a tenuous floristic division into boreal, arid, and long summer forest phases of the holarctic will be made.

Boreal Phase

The boreal or northern subrealm of the holarctic flora is characterized particularly by certain genera of dominant trees that are unimportant elsewhere and by the general association of these trees with a limited set of other species in nearly uniform stands over vast distances (Map 3-2). Secondarily, the majority of plant species of the boreal zone extend throughout much or all of the zone but are essentially limited to it. Except for the few distinct tree genera, however, the plants belong to common genera of the holarctic or even to cosmopolitan groups and no sharp line can be drawn to divide the boreal zone from the more southerly floristic phases. Trees do not extend throughout the boreal land areas, tending to disappear at higher latitudes and higher elevations where cooler, briefer growing seasons are experienced. Beyond the trees the general term "tundra" applies (or "alpine" at higher elevations) which simply means treeless.

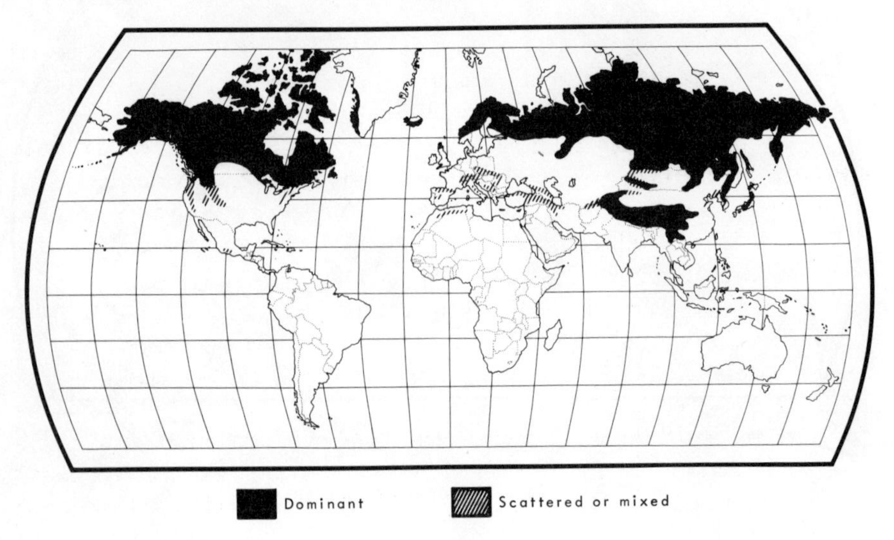

MAP 3.2. Distribution of the boreal flora or cool holarctic.

Trees of the cold northern areas are restricted to six widespread genera: *Picea, Abies, Larix, Pinus, Betula,* and *Populus.* Individuals of *Salix, Alnus,* and perhaps *Sorbus* are also common at higher latitudes where they generally grow as low, bushy forms. That four of the six tree genera are conifers has led to the name, Northern Coniferous Forest, a label which can be misleading. The pines (*Pinus*), birches (*Betula*), and aspens (*Populus*) include numerous other species in warmer zones but spruce (*Picea*), fir (*Abies*), and larch (*Larix*) are nearly limited to the boreal flora with only a few species, as in the Pacific Northwest of the United States, growing in milder climates.

With the trees grow a larger variety of bushes and herbs, prominent among which are several heath genera, blueberry (*Vaccinium*), viburnum, spiraea, juniper, rhododendron, dogwood (of the non-flowering types), sedges, rushes, and grass. Foliaceous lichens like reindeer moss are particularly noticeable in the wetter woodland zones and true mosses are widespread. An irregular topography resulting from recent glaciation together with permafrost combine to produce a great deal of bog land divided into generally small units, while lichens tend to appropriate soilless upland areas. Where the trees become less dense, the cover of low plants naturally increases to take over completely beyond the tree line. Not all of the woodland bushes and herbs extend into the tundra while prostrate forms particularly of birch and willow form a part of the tundra. The variety of species continues to decrease toward the far north until only a few scattered types survive in ice free areas in the most extreme environments.

The treeless areas are a mosaic of herbs, thickets, and bogs including a broad transition zone with tree "islands." The tundra proper is dominated by sedges including cotton grass but it also contains true grasses and a variety of herbs and prostrate woody plants that make a colorful floral display in the brief summer. It is important to realize that virtually the entire species content of the tundra grows within the forest zone as well, so the tundra is floristically only a subtype and not a distinct unit.

Lowland boreal flora can be separated into two or perhaps three provinces. The majority of the genera are divided into distinct but closely related species in North America compared to Eurasia. A further division occurs between Europe and Siberia but often with a gradual transition from west to east which may be incipient speciation over such a broad territory or may be advanced hybridization. Even within North America there are differences from east to west but seldom is this more than variation of species content.

Distinct species often occur in extensions and outliers of the boreal flora in lower latitude mountain areas. Fir extends as far south as Honduras while the Himalaya and the Alps are host to a wide range of boreal genera. The smaller and more remote highlands naturally display a more limited selection of the flora. The distribution of larch can show the nature of the geographical variation of the boreal assemblage. In the first place, there are three higher latitude species, tamarack or American larch across North America and Russian or Siberian larch which grades eastward into Dahurian larch on the Eurasian landmass. In the Rockies is the subalpine larch, in the Alps the European or common larch, in the Himalaya the Sikkim larch, east of Tibet is the Chinese larch (and also the rare Master's larch), and finally there is the Japanese larch. Western larch grows within a warmer zone in North America and there is no larch in the Caucasus. (The closest relative of *Larix* is *Pseudolarix* which grows in subtropical forests of China.) Each mountain region is as floristically distinct as either lowland boreal province but it is not generally given equal treatment because of the limited area involved.

Long Summer Forest Phase

The bulk of the holarctic flora is contained within the robust middle-latitude forests, particularly in the wetter parts (Maps 3-3). In North America and eastern Asia, this long summer forest subrealm of the holarctic has a north-south range from about 50° N. to 25° N. at low elevation and in highlands to about 10° from the equator. In Europe the latitudinal position is displaced to the interval of 60° to 40° and in highlands to 20° with the same position along a coastal strip of western North America. Local divisions within the forest are made which reflect moisture and temperature gradients with a tendency for floristic reduction toward peripheral locations and more severe environments.

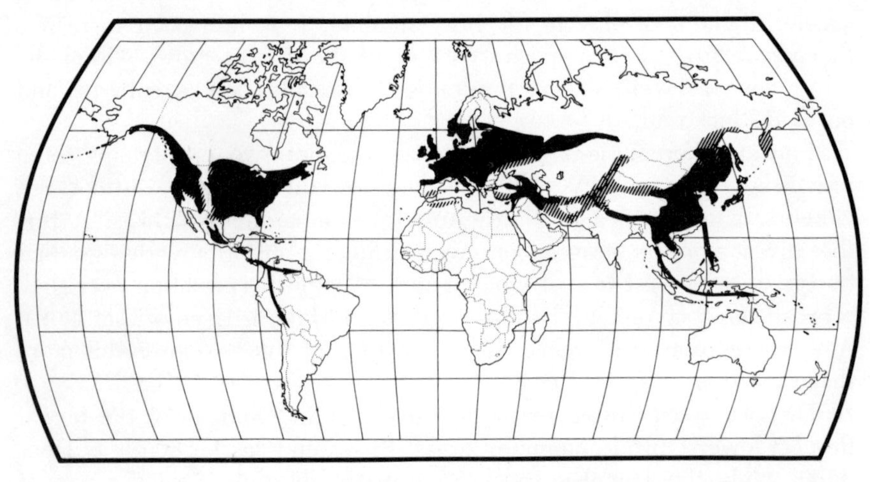

MAP 3.3. Distribution of the long summer flora of the holarctic.

Humid holarctic forests center around several major and lesser nodes of favorable conditions for plant growth where a variety of habitats are found, particularly Europe, China, eastern North America, and western North America. All four of these have a great deal in common at the generic level and can be compared in terms of a table of their important tree genera. Twenty-two tree genera occur in all of these regions and another 39 are shared between the Old World and the New World. Only in eastern Asia are there more than a few tree genera not represented elsewhere in the holarctic flora. Just a few holarctic genera such as *Ilex* (holly), *Diospyros* (persimmon), and *Persea* (bay or avocado) are as well-developed outside of the holarctic as within it. On the other hand, the regional dominants are also among the most widespread genera. Oak forests occur in all parts of the holarctic as do the generally secondary stands of pine. Maple and birch are more typical of cooler and wetter areas all around the northern hemisphere while hemlock and basswood are only a little less prominent in the same areas. Many common understory and old field trees are also familiar from west to east including dogwood, hazel nut, redbud, hackberry, hawthorne, and juniper not to mention the ubiquitous poplar and willow. Throughout the holarctic long summer forests, the majority of the trees of one region are of the same genus as those of any other region. Many herbs are also widespread but largely holarctic in distribution such as the pinks, the buttercups, the poppies, the mustards, and the carrot family. Others are cosmopolitan for the most part.

The strong generic uniformity of the middle latitude forest is shared among four distinct major provinces, already mentioned, and various lesser areas, all with their own particular set of species. Europe is floristically the poorest of the four major provinces characterized by mixed forests of oak, beech, and hornbeam or on warmer south slopes with oak, chestnut, box, and other types. Less tolerant trees such as ash, birch, and pine dominate along with oak to the west and north. Eastern Asia has the richest middle-latitude flora and is the only holarctic forest province in contact with tropical rainforest. A large minority of the eastern Asiatic tree genera are unknown elsewhere. The culmination of holarctic forest is the so-called mixed mesophytic forest of the Yangtze Valley with outliers in southern Korea and middle Japan. This largely deciduous forest is said to lack any clear dominants because of the large number of kinds of trees present, with perhaps more than sixty genera being represented in the canopy alone. However, considering the generally highly disturbed state of these forests, division of the trees into successional and habitat variations as well as local and scattered species can reduce the complex considerably. The most characteristic types are oak, a relative of hickory (*Pterocarya*), and sumac. Other important types are maple, chestnut, ash, walnut, China fir (*Cunninghamia*), magnolia, Douglas fir, and hemlock. There are nearly as many important endemics as there are genera known outside Asia. The composition of the East Asian forest diminishes northward into colder and drier areas, often with different species appearing. The northern forests are dominated by oak with varying amounts of birch, maple, basswood, ash, hornbeam, and pine. To the south, broadleaf evergreen trees become predominant, particularly species of oak and its close relatives but also members of the laurel and magnolia groups. A wide variety of other genera are present.

The two major provinces in North America resemble Europe and Asia in many ways but have some peculiarities also. The forests of western North America, like Europe, have a limited variety of trees but are unique in the relative dominance of evergreen conifers. The most robust forests are controlled by hemlock while Douglas fir is the most widespread dominant. Species of true fir and spruce descend into the lowland forest along with pine, maple, tan oak, and, in California, the redwood. Drier forests see the Douglas fir mixed with pine, oak, madrone, and incense cedar. Eastern North American forests also center on a mixed mesophytic forest as in eastern Asia. The most common trees here are maple, tulip tree (*Liriodendron*), beech, basswood, oak, chestnut, and buckeye; all also are present in Asia but in different proportions. To the north this forest gives way to a less varied stand of maple, beech, birch, and hemlock with pine and basswood. The drier areas to the west are dominated by oak and hickory with pine added to the south; but in wet areas to the south a more complex forest develops with broad-leaved evergreens prominant, particularly magnolia but also oak, bay, and holly.

In addition to the four major forest provinces, there are several lesser regions mostly in mountainous areas. One of these is the area between the Black and Caspian Seas which has strong affinities with Europe and is famous as the probable source of many cultivated fruit and nut trees. The forests of the Himalaya slopes are distinct and rich, outstanding for their many rhododendron trees in addition to numerous other widespread tree genera. The highlands of Mexico and Central America share a little each with eastern and western North America along with the world's greatest variety of pines. Lesser areas of mountains, such as the Atlas Mountains, also have some individuality.

Arid Phase

The trees of the arid phase or subrealm of the holarctic are practically all of genera important in humid areas as well, but the shrubs which assume a greater role include some types that center on dry regions (Map 3-4). The area referred to here is developed within the same latitudes as the long summer forest phase although parts of the tundra are also arid but are floristically distinct. Within the middle latitudes, a three-way division into woodland with more trees, desert with more bushes, and grassland without woody plants can be made.

Drier interiors of the holarctic long summer zone have a reduced complement of trees differing only in species from their forest relatives. Prominent are oaks, pines, junipers, and plums, with poplars and willows in bottomlands. Among the trees and beyond their range are widespread shrub types such as sagebrush (*Artemesia*), gooseberry (*Ribes*), sage

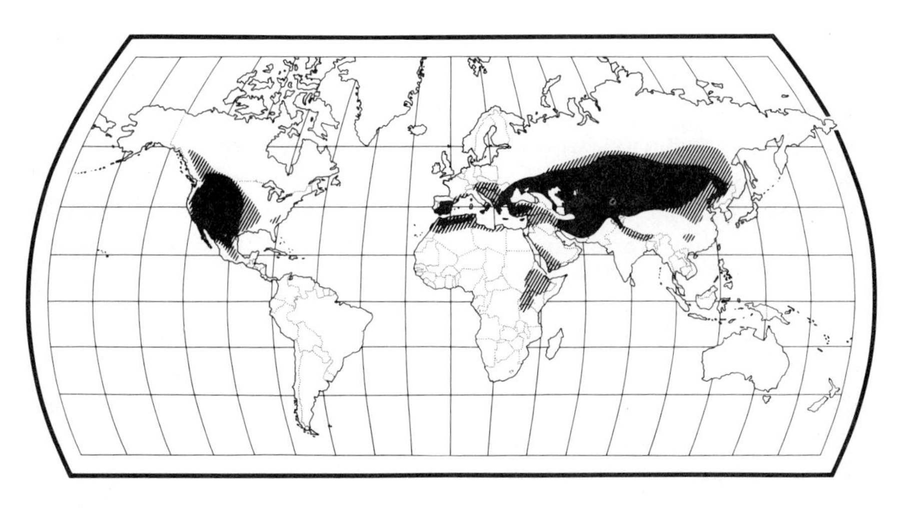

MAP 3.4. Distribution of the arid flora of the holarctic.

(*Salvia*), and scrub oaks. Various salt bushes, particularly of the genus *Atriplex,* become important in the driest areas. Herbs, many of world-wide genera, grow among the woody plants including grasses (*Andropogan, Stipa, Sporobolus, Agropyron,* and *Boutelona*) and forbes such as clover, sunflower, aster, thistle, goldenrod, and many others. Endemic shrubs and herbs also occur. The great grasslands differ only in that the woody plants have essentially been removed. The distinctiveness of the arid-phase flora is contained primarily in the salt plants and the endemics or, otherwise, only in species differences and the absence of most forest tree genera.

There are really only two provinces in the arid part of the holarctic although each can be subdivided into numerous basins, among which are the tropical bordering areas with mixed floras penetrated by tropical plant groups. The dry lands of inner Asia are particularly monotonous with grass steppe or wormwood brush (as the Old-World sagebrush is called). Very characteristic, however, are the regional genera including particularly *Haloxylon* and *Tamarix,* the latter of which has become widely naturalized in North America. In North America, important regional genera are *Purshia* (bitterbrush), *Sarcobatus* (greasewood), *Ceanothus,* and *Manzanita.* Unlike three of the four forest provinces, the two arid provinces are widely open to contact southward for floristic exchange with the comparable tropical floras.

Contact Zones

The holarctic floristic realm is bordered on the south by other realms and some interchange occurs. The penetration of holarctic elements beyond the zones generally dominated by them has been largely along high mountain chains where their tolerance of cold is of decisive importance. The greatest spread into tropical highlands can be seen in Malesia where rhododendron, oaks, and tan oaks are numerous into New Guinea while pine and yew among other types are well established beyond the mainland. The southeast Asian highland flora is really a continuous extension from the north. A similar extension occupies the highlands of Mexico and beyond while oak, alder, and walnut, for example, are important in parts of the South American Andes mountains. The Sahara has been an effective barrier against the holarctic flora but the highlands along the Red Sea have formed a limited corridor. One species of juniper, for example, is significant in the east African highland flora. The Sahara itself is host on the northern side to a limited mixture of *Atriplex, Tamarix,* and *Artemesia* among a much larger variety of non-holarctic plants. It is notable that, for the most part, in spite of the examples just given, the holarctic flora terminates almost abruptly beside other floras, particularly on the flanks of mountains. The boundary of the holarctic flora is perpendicular to environmental gradients, that is, parallel to the limits of more or less tropical conditions.

neotropics

New-World tropics, Mexico to central Argentina, is the area designated as the neotropics, having much to contrast with the other realms in terms of flora. Both to the north and the south the neotropical flora ranges sparingly beyond the tropics where it soon gives way to different plant groupings. A pronounced distinction can be made between the humid and the arid parts of neotropics so that two contrasting phases or subrealms are recognized.

Humid Phase

The vast wet forests centering on the Amazon Basin probably consist of the richest assemblage of plant species in the world (Map 3-5). Heavy precipitation along coastal mountain fronts carry the rainforest nearly 25° from the equator in Mexico and 30° in southern Brazil. Trees are the most important life-form of this forest which displays a remarkable regional uniformity in spite of its local complexity. Not surprisingly, however, it is somewhat reduced and altered in composition up mountain slopes and into the drier margins. Entirely different plants replace the lowland rainforest, sometimes abruptly, higher up in the mountains or where severe dryness occurs.

Several thousand genera of plants grow only in the neotropics, a sizeable part of the total world flora, which justifies its separation from

MAP 3.5. Distribution of the humid flora of the neotropics.

other tropical floras. Some of the common and economically important American trees include the Brazil nut (*Bertholletia excelsa*), kapok (*Ceiba*), balsa (*Ochroma*), cedar (*Cedrela*), rubber (*Hevia*), chicle (*Achras*), quinine (*Cinchona*), and many others. From the smaller American trees have come cacao (*Theobroma*) and papaya (*Carica*). Among the characteristic herbs (often epiphytic) and vines are the vast array of bromiliads from pineapple to Spanish moss, the cannas, the pokeweed group, begonias, bougainvillea, and passion flower. On the other hand, there are also many prominent groups of wide distribution as well, such as the Brazil wood (*Caesalpinia*) which was the first tree to be exploited in America, banyan (*Ficus*), many varieties of legumes

(*Leguminosae*), Bombax, Ilex, numerous orchids, and even mahogany which also occurs in Africa. This mix of local groups with groups extending through several floristic realms tends to distinguish the tropics in general from the cooler areas, particularly the holarctic where a large majority of the genera are unimportant outside of their one realm.

In addition to the immense Amazon forest proper, the humid subrealm of the neotropics includes several outlying regions none of which really merits the name of province. The most distinct area is the West Indies where scattered stands of wet forest occur on most of the islands generally enduring considerable seasonal drought. In Middle America one could distinguish a drier west coast forest and a wetter one on the eastern side. Finally, cut off by a drier and higher zone, itself distinct, is the coastal forest of Brazil south of the eastern "buldge." The distribution of mahogany species illustrates the relationships. The most widespread species (*Swietenia macrophylla*) completely encircles the Amazon Basin (but does not enter the wet central lowland) and continues along the eastern side of Central America and Mexico. A second species (*S. mahogoni*) is restricted to the West Indies and southern Florida, while a third (*S. humilis*) extends along western Mexico and on to Costa Rica. There is no mahogany in coastal or highland Brazil, but species of plants in each of these areas are usually different from the other regions. (The mahogany of Africa is put into a different genus on shaky and probably untenable grounds). The widespread savannas in America are partly related in flora to the humid forests. Galeria forests and other forest fragments intrude into the grasslands, and these are comprised in the wetter areas of rainforest extensions and in drier areas of seasonal forest groves. The trees of the open savannas, fewer in variety than in the forests, are largely shared with the forest areas, often occurring as successional elements or growing on difficult sites. The legumes are particularly prominent savanna trees. It is notable that the savannas are located within the drier forest zones, western Central America, the Indies, northern South America, and the Brazilian highlands. Although the wetter parts of Central America are virtually continuous with the Amazon forests, there are differences, as the fact that chicle is confined to Central America where the rubber tree is absent.

Arid Phase

The flora of drier conditions within the neotropics stretches from southwestern United States to central Chile and Argentina with an extension into eastern Brazil (Map 3-6). The core of this subrealm lies in the deserts, but less arid areas of probably greater extent are also included. As one approaches the forests, floristic elements most closely related to the humid part of the neotropics begin to appear. The fact that

MAP 3.6. Distribution of the arid flora of the neotropics.

the arid phase of the neotropics is largely independent from the humid phase suggests that it could easily be set aside as a distinct realm but these two phases being so intimately intertwined within the area of Latin America would make such a division awkward.

The most unique and visually dominant element of the arid subrealm is cactus of many types. The general categories of prickly pear (*Opuntia*) and organ pipe (*Cereus*) are found almost everywhere within the subrealm. The yucca-like plants of the lily family are less ubiquitous but no less striking. The most common plants are nondescript spiny bushes and trees particularly of the legume family including mesquite (*Prosopis*), the worldwide *Accacia, Mimosa, Caesalpina,* as well as palo verde (*Cercidium* and *Parkinsonia*), which are more typically American, and quite a few others of more local occurrence. Among widespread and characteristic desert shrubs are the creosate bush (*Larrea*), salt bush (*Atriplex*), and bur sage (*Franseria*). Other shrubs including the legumes have desert forms; some are local and others such as *Ephedra* are nearly worldwide in dry climates. Groves of fan palms occur here and there, often around springs, from one end of the arid zone to the other. Herbs are generally of extensive groups such as composites, nightshades, cucurbits, and grass.

There are two main provinces of the arid neotropics, one centered on Mexico and the other in South America at the same latitude of the southern hemisphere. In spite of being rather isolated from one another these two areas have so much in common that parallel development has been suggested. However, there are scattered drylands all through Central America, the West Indies, and the highlands of northern South America. At the same time, both are compartmentalized by the many major mountain ranges into numerous subprovinces. In wetter areas there are forests and woodlands filled with thorny trees, cactus, and in some areas palms. Feral citrus is found everywhere. The warm deserts of Mexico and adjacent United States have the peculiar ocotillo (*Fouquiera*) and in Baja California is its even more peculiar relative (*Idria*). Agave is very characteristic of deserts and woodlands of the same province. In South America, relatives of sumac are important. The pepper tree (*Schinus*) grows all along the Andes and to Uruguay and is even common in Mexico where it is apparently naturalized. *Schinopsis* includes the valuable quebracho of the Chaco, a name also given to the unrelated *Aspidosperma*. Woody composites are of interest in the dry areas around the Andes. A lesser outlying dry area is the Caatinga of northeastern Brazil where the general character of the flora is much the same as that of other dry parts of the neotropics but with certain differences. Perhaps because of its separation there are more unique genera or, more particularly, genera otherwise associated with nearby wet forests and not with dry areas elsewhere. Mesquite and pepper tree are missing here. The widespread palm *Copernica* has a species in the northern Caatinga which produces carnauba wax (*C. cerifera*). Further west an endemic palm, the babaçu (*Orbignya martiana*), forms extensive groves and its oily nuts are the local staple.

Contact Zones

The major contacts of the neotropics with other floristic realms involve the dry phase. There are two such areas, one in the north and the other in the south. Here lowland areas with gentle environmental gradients yield the greatest possibilities for intermingling. Along mountain slopes the floristic change is almost abrupt. Also along mountains as on the eastern side of Mexico and the eastern sides of the Andes outliers of wet forests from extra-tropical floras do have contact with wet neotropical floras.

In North America the holarctic pine-oak woodland usually lies on the cool side of the neotropical flora where these come together; usually this is upslope where mesquite and cactus give way in a very short distance to pine and oak. The greatest mixture occurs in the lowland deserts, particularly in the Mojave desert but also in west Texas. A narrow strip of moist holarctic forest with sweet gum, walnut, and oak lies above a strip of neotropical rainforest in eastern Mexico south to Nicaragua. A few neotropical elements have penetrated into the holarctic realm itself. Small cactus are a minor element right through the western North American arid lands and a cactus extends as far as Long Island in sandy and shallow soils. Mesquite overlaps broadly in Texas and New Mexico. Several tropical plant genera extend into southernmost United States and a few even more deeply, such as persimmon (*Diospyros*), bay (*Persea*), and ironwood (*Cyrilla*). Holly and *Celtis* which are largely tropical genera, appear to be well integrated into the holarctic flora as well, as are a number of herbaceous groups. Notable, nevertheless, is the limited variety of the northward penetration of neotropical plants.

The southern margin of the neotropics consists of arid conditions stretching from the Pacific to the Atlantic at about 35° S. latitude. To be sure, Antarctic plants do occupy parts of the Andes further north and dominate a broad area of highland southern Brazil. The deserts of Patagonia see a few straggling elements of the neotropics, particularly legumes and composites along with grass, extending into the middle latitudes. On the southern, wetter side of the desert the Antactic flora soon becomes dominant but a mixture of elements is found, particularly in middle Chile, over considerable territory.

palaeotropics

The area here being considered under the term palaeotropics is limited to Africa and lowland southwestern Asia although the term is often extended to include Malesia. With the exception of the projection into western India, this is perhaps the most compact of the four floristic realms. It may also be the least rich in variety. There are two distinct subrealms which can be recognized; in the center lies a humid complex while both to the north and to the south are arid nuclei. Between these poles lie a series of transitional zones of fairly distinct character.

Humid Phase

Across the center of Africa from the Guinea coast through the Congo Basin grows a luxuriant rainforest (Map 3-7). Unlike the case in South America, the fully developed rainforest forms a rather narrow belt on either side of which lies a seasonal forest. For its part, the seasonal forest is widely replaced by savanna as, in fact, are also the margins of the rainforest. In the savannas, remnants of the forest reveal the relationships.

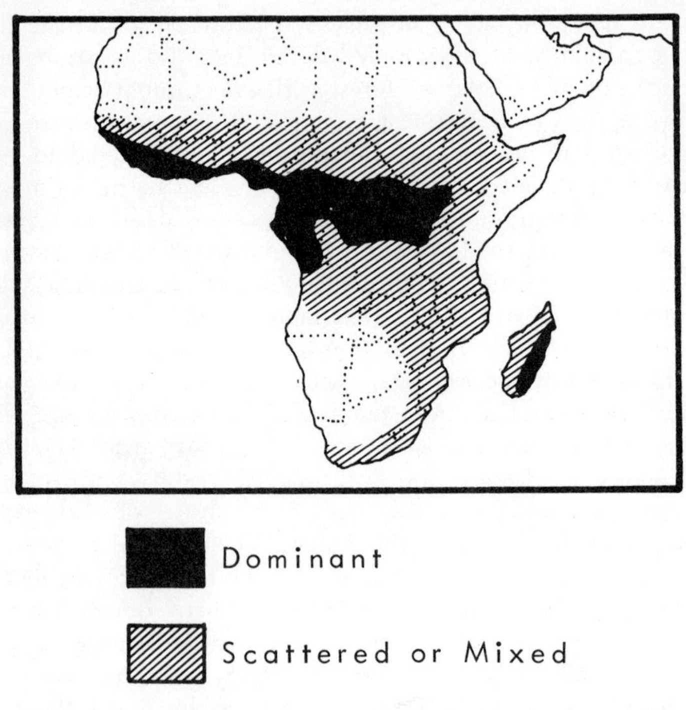

MAP 3.7. Distribution of the humid flora of the palaeotropics.

Among the thousands of rainforest tree species are few that are well-known elsewhere and because a rainforest has no real dominants, the flora is difficult to characterize. Import genera include *Iophira, Cynometra, Antiaris, Chlorophora, Tarrietia,* and *Drypetes.* Some widespread groups are also represented including banyan (*Ficus*), ebony (*Diospyros*), kapok (*Ceiba,* some think it is introduced), and *Albizzia.* The oil palm (*Elaeis quineensis*) is an important native staple and these forests are also known for cola and for coffee (the commercial form comes from a highland species). In drier parts of the rainforest other kinds of trees become important, many of which are of very widespread genera such as *Terminalia, Celtis, Sterculia,* and *Triplochiton.*

The seasonal forests contain many elements that are related to the rainforest flora but are not of the same species. The three most dominant genera are *Isoberlinia* of the north and *Brachystegia* and *Julbernardia* of the south. Secondarily in both north and south are found species of *Terminalia, Combretum, Monotes,* and *Annona* (custard apple), as well as other more local or less common trees. The famous baobob (*Adansonia*) overlaps the forest zone and drier areas. Very coarse and tall grasses grow close to the rainforest but these give way to finer and short grasses within the main part of the savanna zone. A great variety of poorly studied forbes of world-wide affinities grow among the grasses.

The moist palaeotropics can be subdivided into several major parts, sometimes given the rank of provinces, and several smaller areas. The main part of the rainforest is the Congo Basin. The coastal strip in west Africa is not quite continuous with the main forest and has somewhat fewer species. Strangely, the rainforest of Madagascar has significantly more variety than that of the Congo. Small rainforest remnants are also found along the coast in Kenya and Tanzania. Savanna dominates the landscape north of the rainforest both in the seasonal forest zone and beyond in the arid zone. All of this belt of open country is referred to as the Sudan but as a distinct floristic province it is questionable. South of the Congo Basin the landscape is dominated by a great block of the so-called myombo woodland from the Angola highlands to the Indian Ocean. The characteristic tree genera of this are *Brachystegia* and *Julbernardia.*

Arid Phase

The arid part of the paleotropics is the greater portion extending solidly between 15° and 35° in latitude both north and south and to the equator around the horn of East Africa (Map 3-8). It is particularly characterized by the woodlands even though there are great areas of (almost plantless) desert in the Sahara. A considerable floristic overlap occurs with the moist palaeotropical flora where they meet. By far the most conspicuous plants are the acacias some of which have most persuasive thorns. Savanna alternates with the woodlands in all degrees of variation from grassland to tree cover. There are two distinct provinces, a northern and a southern, with connections through the rift valleys of East Africa.

The northern part of the arid subrealm extends from the Atlantic Ocean to western India and includes the Sahara, Arabian, and Thar deserts as well as much of the Sudan. The woodlands contain several notable trees including myrrh (*Commiphora*), olive (*Olea*), and a well-branched palm (*Hyphane*). Acacia continues to be important in the desert zones as well as *Astralagus* and a variety of shrubs among which are a few holarctic elements. There are many forbes and kinds of grass, much of which grow only briefly when rains occur. The driest parts of Burma have a veritable outlier of the flora of north Africa with woodlands of thorny acacia and prominent euphorbs.

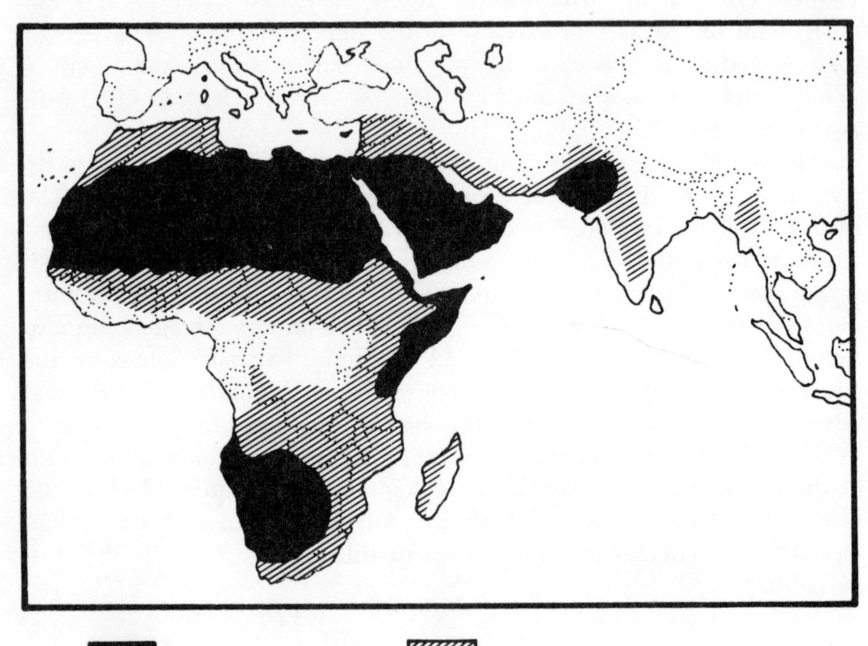

MAP 3.8. Distribution of the arid flora of the palaeotropics.

South of the equator, the general nature of the arid flora is the same as to the north. In southern and eastern Africa arborescent *Euphorbias* are prominent, paralleling the forms of cactus in the neotropics. *Aloe,* resembling *Agave,* is also characteristic. The coastal desert of South Africa has particularly abundant succulent plants including the greatest concentration of *Mesembryanthemum.* The curious *Welwitschia* occurs here.

The southwestern corner of South Africa deserves special mention for its flora. The great concentration of local endemics in this small area has led some plant geographers to elevate the flora to the status of an independent realm. There are hundreds of species of *Erica* (heaths) here, although this genus has spread to many continents. Numerous other plants of many different families have developed local members with a similar (ericoid) form within the "cape macchia." Apparantly the small area around the southewestern tip of Africa with its isolated bit of moist extra-tropical climate represents a refuge of a flora which, if developed over a larger area, could easily merit distinction as a subrealm. Despite its strong distinctness, however, it contains important reciprocal relations with the rest of the African flora as well as a significant austromalesian element which will be discussed later in this chapter.

Contact Zones

The palaeotropics are in contact with other floras today only on the north where a very long border zone exists; some kind of contact seems to have existed with Australia at one time. Very few palaeotropical elements extend beyond the limits of its territory. The northern boundary, for all its length, everywhere abuts mountains up the slopes of which rather considerable differences in conditions exist. Furthermore, the most accessible boundary zone passes through the Mediterranean Sea. It is around the Mediterranean that the greatest interchange of floristic elements has occurred. The commercial olive and the edible fig (*Ficus*) are palaeotropical standard bearers in Europe (in effect) even though both these genera are rather widespread. The northern migration of heath was mentioned before. The path of migration may be obscure but the olive genus also extends as far as New Zealand. *Adansonia* is also found in Australia. The examples are few and the palaeotropics remains well differentiated from the other floristic realms.

austromalesia

The floral area being grouped together here as austromalesia comprises far-flung and a very fragmented assemblage of land areas, islands and corners of continents. These are not customarily so grouped, although there is every bit as much reason for doing so as for grouping the various parts of the neotropics or of the palaeotropics. The area begins with peninsular India, continues through southeast Asia, Indonesia, Australia, and the major Pacific islands ending at the southern tip of South America. A great many interesting relationships exist across this far-flung domain and three well-marked phases immediately present themselves. The first is humid and tropical while the second, involving mainly Australia, is arid. The third is cool and is called Antarctic.

Humid Tropical Phase (Malesia)

The first subrealm of the austromalesian flora is entirely tropical within which there is very little truly arid land (Map 3-9). Included is the

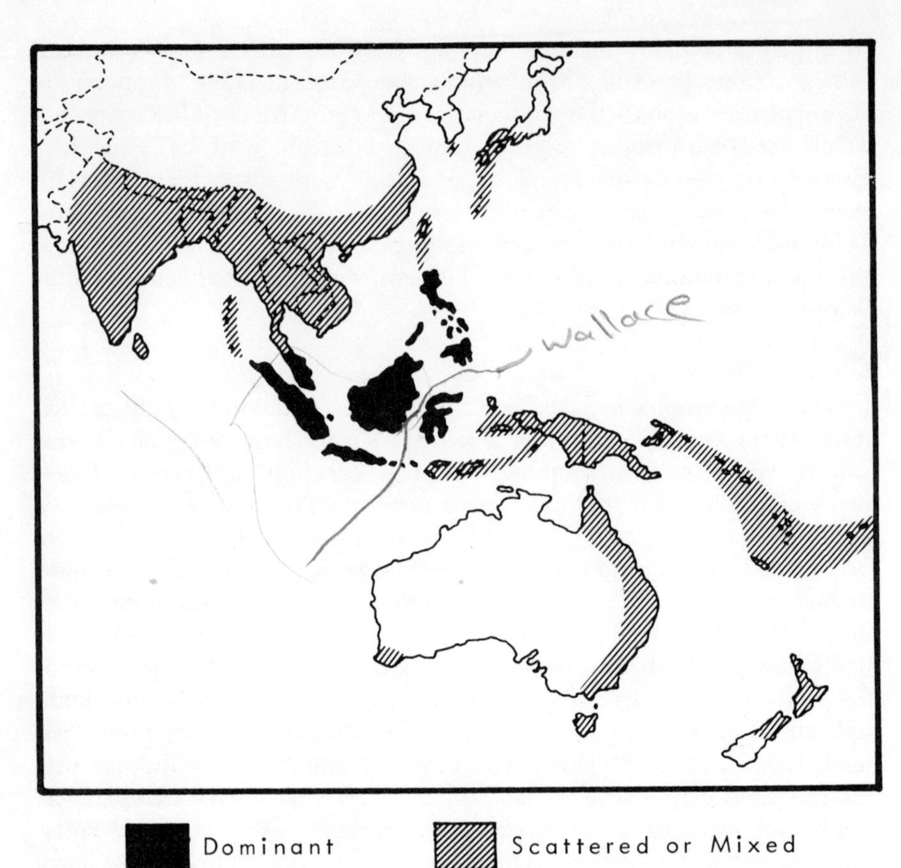

MAP 3.9. Distribution of the malesian flora or humid austromalesia.

classical area of Malesia itself together with peninsular India and the intermediate regions. Such arid areas as the Thar desert and central Burma are dominated by African elements and are not really a part of this phase. There are extensive seasonal forests in India, Burma, and southeast Asia whose flora is derived from the richer rainforest flora. Between these seasonal forests and the wettest rainforest are zones of intermediate complexity, and much species overlap so that floristic divisions are largely artificial.

The Malesian rainforest, although rich in diversity, has some very distinct floristic elements. The family Dipterocarpaceae is both predominant in the area and largely restricted to it. There is also an important conifer element not found in any other lowland rainforest. *Agathis,* a broad-leaved conifer, is a leading emergent in many areas of the rainforest. Other parts are called dipterocarp forest, but members of Podocarpaceae (*Podocarpus, Decussocarpus, Prumnopitys*) are widespread. There are numerous other essentially endemic genera which totally encompass nearly half of the genera present including the emergent, *Altingia,* and the sago palm, *Metroxylum.* Among the more widespread kinds of tree are banyan (*Ficus*), *Diospyros, Albizzia, Dalbergia, Hibiscus,* and *Sterculia.* Bamboo is particularly evident in disturbed areas and the wild banana is found here. Perhaps the coconut had its origins in Malesia as well.

Seasonal forests lying on the northern margins of the rainforest are largely derived from its flora. The two most common trees are sal (*Shorea*), a dipterocarp, and teak (*Tectonia*). Other important genera are *Xylia, Pentacme,* and the widespread *Terminalia, Lannea,* and *Adina.* Locally are found such types as *Dipterocarpus, Sterculia, Acacia, Pterocarpus, Albizzia, Diospyros,* and many others. The zones of overlap with rainforest flora are broad.

The humid austromalesian flora can be said to have two main provinces (or three if one wants to call the Pacific Islands a separate province). One of these is south Asia with a major division between India (sal forests) and southeast Asia (teak forests). The other province is Malesia which is further divided into a western (Asiatic) and an eastern (Australian) subprovince, separated by the famous Wallace Line (See Map 3-13). The western part is particularly dominated by dipterocarps and has greater affinity with adjacent Asiatic floras. The eastern part has fewer dipterocarps and a mixture of Australian genera. Prominant in the east are emergent *Araucaria.* In spite of this division into west and east, a large part of the flora is continuous from one end of the rainforest to the other.

Outlying fragments of humid tropical austromalesian flora are found in several areas. Peninsulas extend to the foot of the Himalaya and along the south coast of China. Formosa, Ceylon, and the western Ghats of India are included. Furthermore many continental islands, not just the Solomons, but beyond in New Caledonia, New Hebrides, and Fiji are a part of the same flora. Admittedly modified but nevertheless predominantly derived from the same source are floras of the Pacific islands as far away even as Hawaii. Finally the rainforests along the northeastern coast of Australia belong to the humid tropical austromalesian flora.

Arid Phase (Australia)

The great dry bulk of the Australian continent is host to a distinctive flora which extends to a limited extent into southern Indonesia, southern New Guinea, and the drier parts of New Caledonia (Map 3-10), More than any other major floristic region it is as much in the tropics as it is beyond. The dominant floristic elements are not divided between the wetter and drier extremes although deserts, grasslands, and savannas occur as well as woodlands and forests. Particularly striking is the agressiveness of Australian plants where they have been introduced into comparable climate areas in other parts of the world.

Far more than any other genus, *Eucalyptus* dominates the flora of Australia penetrating all the major habitats. Also common is *Acacia,* a worldwide genus, but in Australia and nearby lands represented for the most part by a phylloidous subgenus whose expanded leaf bases replace the usual much-divided leaf blades and resemble the eucalypt leaf in form. Other associated plants are *Casuarina, Callitris, Cycas, Banksia, Hakea,* and grass trees (*Xanthorrhoea*), all of which are re-

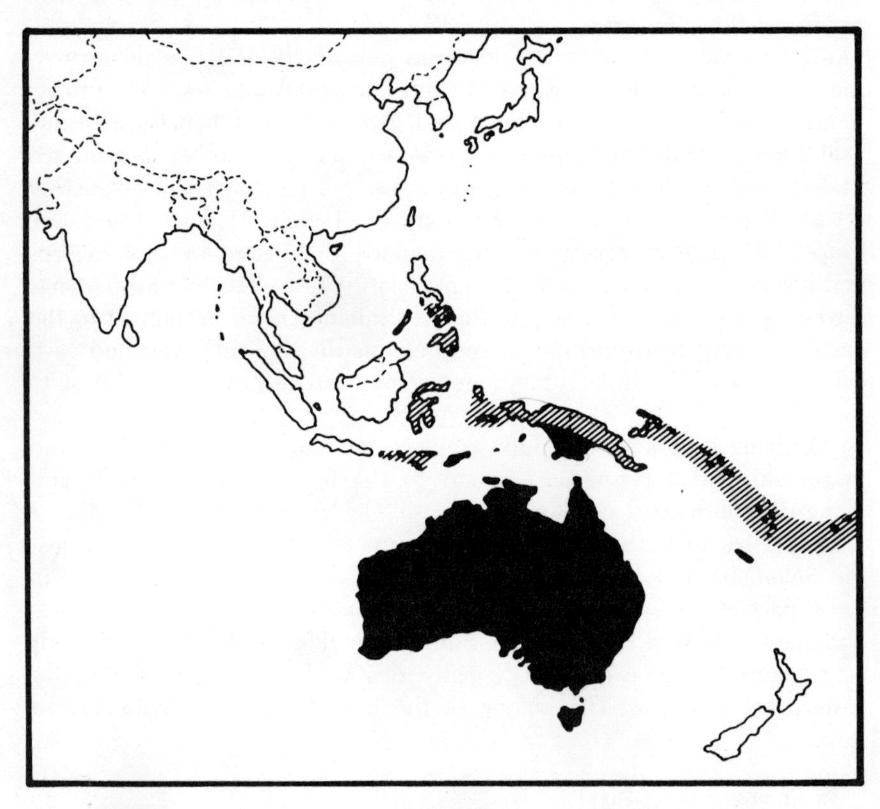

MAP 3.10. Distribution of the Australian flora of arid austromalesia.

stricted to the austromalesian realm, some only in this arid phase, or, as in *Casuarina* and *Cycas,* extending into the humid tropical parts. More local within the region are *Melaleuca* (paper-bark) and *Atriplex* (saltbush). Several tough grasses have a role in Australia including tussock grass (*Poa*), Mitchell grass (*Astrebla*), and spinifex (*Triodia*).

Only one compact province comprises the arid phase of the austromalesian realm. Involved are most of Australia, nearby areas immediately to the north, and parts of New Caledonia. Only an interrupted fringe of eastern Australia would be excluded. Elements of this flora can be found as far as the Philippines and Hawaii while massive introductions have been made throughout the warmer parts of the world.

Cool Phase (Antarctic)

The distribution of the Antarctic flora is the most unusual of any comparable flora in the world (Map 3-11). The primary areas are Tasmania, New Zealand, and southern Chile. In addition, however, the Antarctic flora is predominant in the highlands of New Guinea, New Caledonia, parts of the Andes, and southern Brazil. Elements are found even further afield, not to mention fossils in Antarctica proper. Most of this flora is associated with a cool, moist environment but to a limted extent it occupies drier zones in South America and moderate elevations within the tropics.

The dominant genus of the Antarctic flora is *Nothofagus,* the southern hemisphere beech. Most of the other important genera are found on both sides of the Pacific including: *Araucaria, Podocarpus, Drimys, Weinmannia, Laurelia, Prumnopitys, Libocedrus, Eucryphia, Myrtus,*

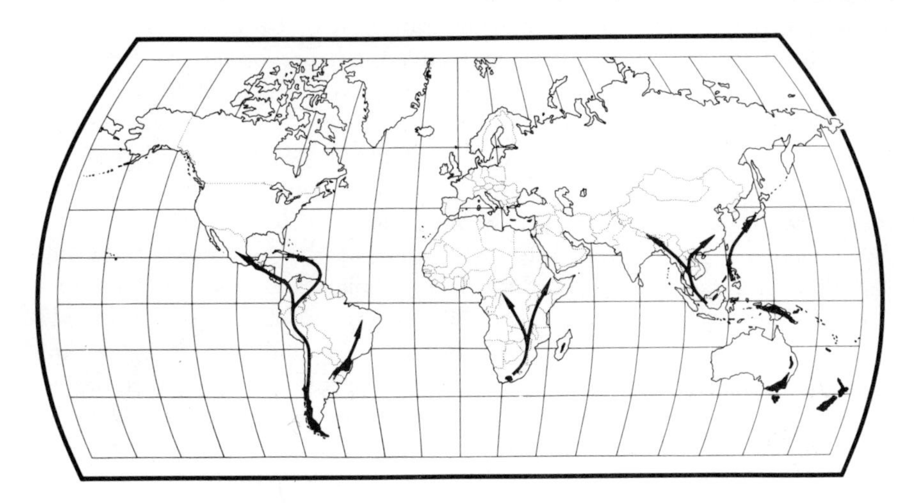

MAP 3.11. Distribution of the Antarctic flora of cool austromalesia.

and *Dacydium*. Half of these are conifers. With the exception of *Myrtus*, these genera are centered on the Antarctic flora as are many lesser plants including *Fuchsia* and the conspicuous tree ferns. On the other hand, only a limited number of extraneous genera penetrate the southern forests.

There are two main divisions of the Antarctic flora with the South Pacific between them. The western region is scattered across lowland Tasmania and New Zealand and then elevated somewhat in New Caledonia and Queensland reappearing again in the mountains of New Guinea. Although the bulk of the flora has representatives on the eastern side of the Pacific, some prominent genera such as *Phyllocladus* occur only in the west. On the other hand, this flora broadly merges with the lowland rainforest and extends along with it as far as Japan and India, particularly such genera as *Podocarpus* and *Weinmannia*.

The South American region centers on the southern Andes. The rich flora includes endemic elements such as *Aextoxicon, Guevina, Saxegothaea, Fitzroya,* and *Chusquia* (a solid-stemmed bamboo). Just as the widespread genus *Olea* penetrates as far as New Zealand, *Persea* is at home in southern Chile. The frost-prone forests of southern Brazil are dominated by such trees as *Araucaria, Podocarpus, Drimys,* and *Weinmannia* of the Antarctic flora. In the central Andes, Antarctic elements such as *Podocarpus, Prumnopitys,* and *Weinmannia* grow beside such holarctic genera as *Quercus* and *Juglans*.

Contact Zones

The Austromalesian flora has contacts in several areas. The greatest mingling occurs on the southeast side of Asia. The abrupt face of the Himalaya makes for a narrow transition to holarctic plants there, but in Japan and particularly in China there are broad overlaps. The highlands of southeastern Asia are a complete mixture of holarctic and Antarctic elements with pine and oak growing with *Podocarpus* and *Dacydium*. Humid tropical and arid elements do not extend to South America but the Antarctic elements enjoy in America the most distant penetration into alien realms. Some species such as *Podocarpus* and *Weinmannia* grow through the highlands of the neotropics right into middle America to the borders of the holarctic. Africa is host to a measure of austromalesian genera. The humid tropics share many widespread plants and a few that are just in Africa and Asia. It is the Antarctic elements again that are most unexpected. South Africa has an important minority of Antarctic genera such as *Podocarpus, Decussocarpus,* and *Metrosideros*. This element further extends northward through the highlands, just as the comparable group does in South America, so that a species of *Decussocarpus* is one of the few highland trees of Ethiopia.

Fauna

The world can be divided into several major faunistic realms which have a great deal in common with the major floristic regions but which differ in certain important ways (Map 3-12). Instead of a minor secondary division along Wallace's Line, there is here the greatest single separation of assemblages. The holarctic is divided between a nearctic and a palearctic while South America has no special relation with Australia and

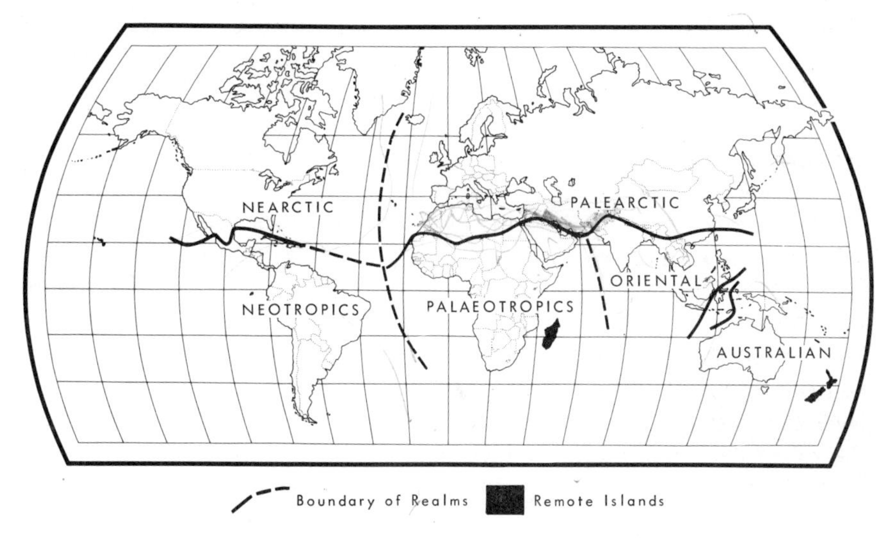

MAP 3.12. Faunistic realms including the larger remote islands.

New Zealand. These differences reflect the separate histories of flora and fauna and their contrasting mobility. For the most part the subdivisions of the faunistic realms parallel the subregions of flora responding to the fact that, within a compact land mass, the moisture and temperature factors and the flora associated with them constitute habitat variations to which animals would naturally respond. The faunistic differences, in addition, are not as profound as those separating adjacent floras, so the division of faunistic realms will not be pursued here.

palearctic

The palearctic is the frost-prone part of Eurasia with the northern fringe of Africa thrown in. The southern boundary lies in the vicinity of the Tropic of Cancer and crosses land areas almost all the way from the Atlantic to the Pacific. In Africa this boundary lies in the Sahara desert while in Asia it is marked by the Himalaya and related mountain ranges. Only a narrow water gap separates the palearctic from the nearctic.

Surrounded as it is by other faunistic realms, the palearctic has no important unique animals. It is distinguished by the peculiarities of its neighbors which will be apparent as each of these is described. There is within the palearctic a particular combination of animals, and these can be identified. The range animals include horses, cattle (bovids), and camels. A wide variety of deer are found along with wolves, cats, rodents and their relatives, and bears. Other common mammals include hedgehogs, pigs, civets, pandas, and mustelids (skunks, weasels, otter). Fish and birds are of many common world groups, notable among which are pheasants, geese, hawks, carp, and salmon. Recently extinct (a few thousand years ago) in this area are mammoths and rhinoceroses.

nearctic

General environment as well as the animals of the nearctic are very much like those of the palearctic; in fact, the nearctic is the New-World equivalent of the palearctic in position, extending south approximately to the Tropic of Cancer. The narrow connection with the neotropics has led to a limited exchange of fauna there, as might be expected, but it must be emphasized that this is only a limited exchange so far as the nearctic is concerned.

The variety of animals of the nearctic can be described in terms of the diferences with the palearctic because the similarities are predominant. There are a few unique forms including the pronghorn antelope, the peccary, and the turkey. The opossum, which is apparently a recent arrival from the neotropics, is widespread. Reptiles are much more abundant than in Eurasia, culminating in the gila monster, the only poisonous lizard. Compared to the palearctic, the nearctic has no hedgehogs, civets, pigs, or camels. The recent lack of camels is notable because they survive in the neotropics as well as being at home in the palearctic. Coons are the common representative of the family to which the Asiatic pandas belong. The horse, the mammoth, the saber-toothed cat, and the giant ground sloth are among the recently extinct large animals.

palaeotropics (Ethiopian)

The palaeotropics or Old-World tropics is limited to Africa and Arabia in and south of the great deserts there. Its only contact lies with the palearctic and, broad though their borderland is, the differences between their animal inhabitants are surprisingly great. As a matter of fact, the palaeotropics has much more in common with the oriental realm which is also tropical but which fails to connect with the palaeotropics where a gap lies in the desert between eastern Arabia and the lowlands of the Indian subcontinent.

As before, the distinctions of the palaeotropics can best be understood by comparison with the palearctic as a base. The animals not found to the north of the palaeotropics include giraffe, hippopotamus, aardvark, chameleon, monkeys, baboons, apes, pangolin, elephant, rhinoceros, loris, and ostrich, as well as numerous smaller animals. On the other hand, tropical Africa lacks bears, deer, camels, beavers, moles, and amphibians with tails. In fact, the bulk of the animals in common with the palearctic are of virtually worldwide types such as cats, dogs, rabbits, rodents, shrews, mustelids, and bovids. Exclusively Old-World types both in and out of the tropics are limited to such animals as horses, pigs, hedgehogs, civets, and few others. The camel, identified today with Africa, was introduced by man. On the other hand, a number of the groups of larger animals have only recently become extinct in northern Eurasia.

oriental

The oriental fauna occupies the tropics of southern and southeastern Asia as far as Wallace's Line. This is the smallest and probably least varied of the faunal regions and for that reason alone can be understood to comprise a lesser range of animals than the palaeotropics which it strongly resembles. On the other hand, its broad and intimate contact with the palearctic is bound to result in some similarities there.

The bulk of the palaeotropical fauna is also found in the oriental realm, including the widespread groups. Notable among those animals not shared in this way are the horse, giraffe, hippopotamus, aardvark, chameleon, ostrich, and baboon. Conversely some northern types, not found in Africa, are bears, deer, and moles. Only a few distinct animals occur in the oriental region, among which are tarsiers and tapirs. This latter is unusual because it also occurs in the neotropics. The oriental fauna is rich in reptiles and birds, notably peacocks and the ancestors of chickens.

neotropics

The area called the neotropics with respect to fauna includes not only the tropical parts of America but also extends to the southern end of South America. The only contact is with the nearctic in Mexico, although the West Indies lie close to Florida.

Fully half of the animals of the neotropics are endemic while many of the rest are of worldwide groups, making this realm the second most distinct in the world, first place going to Australia. One whole major order of mammals is endemic to South America, the members of which are anteaters, sloths, and armadillos. The New-World monkeys with their prehensile tails are distinct from the Old-World monkeys and similar major divisions separate the relatives of South American rodents (capybara, guinea pig, chinchilla, porcupine), bats (including the vampires), marsupials, and the ostrich-like rhea. Other distinct animals include toucans and curassows. The more widely distributed animals of the neotropics include rabbits, rodents like squirrels, shrews, mustelids, bears, cats, camel, tapir, and peccary, as well as numerous reptiles. Bovids, beavers, and moles have not penetrated the neotropics from the north, while mammoths and horses have recently become extinct.

Australia

Australian faunal realm includes Australia proper and the many islands to the north. There are no land connections, the northwestern boundary being Wallace's Line which is placed between Celebes on one side with Borneo and the Philippines on the other, passing further south between Bali and Lombok (Map 3-13). There is some overlap

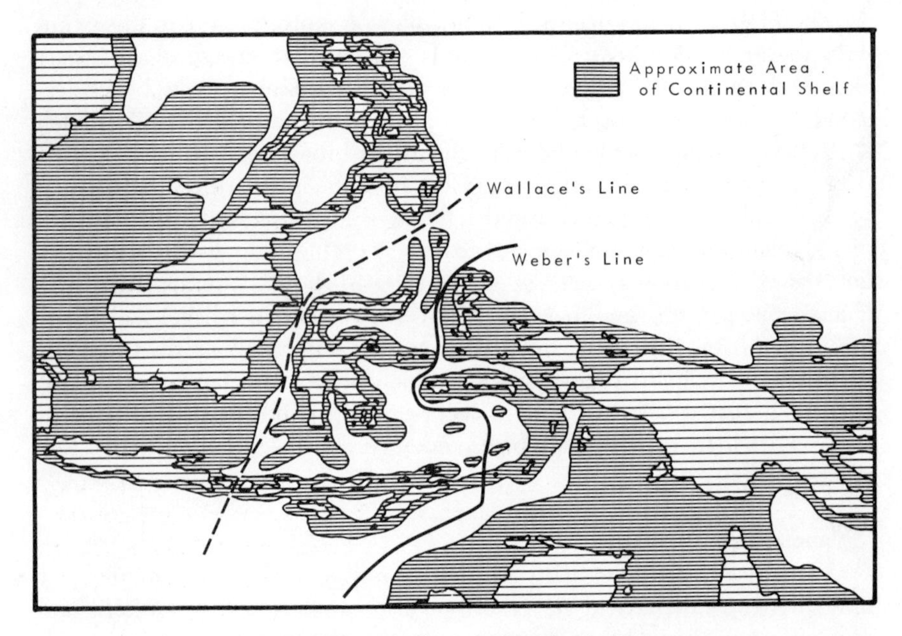

MAP 3.13. Wallace and Weber's lines.

across this boundary line but considering the proximity of islands on either side, the faunal division is remarkably sharp.

The animals of Australia and nearby islands are nearly completely unique if one omits those which fly or swim. Without man probably the only land mammal to invade Australia was one kind of mouse, while Australian mammals are unknown outside of the realm. The native land mammals are primarily endemic marsupials of great variety, the better known of which are kangaroos, wombats, bandicoots, koalas, and a marsupial wolf. Others resemble smaller animals of the rest of the world. Two kinds of egg-laying mammals survive in Australia. Flightless emus and cassowaries have a slight resemblance to the ostrich. The Australian realm has the world's greatest variety of parrots and is notable for birds with elaborate courtships and feather displays such as bowerbirds and the birds of paradise. Pheasants are, however, lacking. Reptiles occur in moderate variety but amphibians are poorly represented, as are freshwater fish, of which virtually all are closely related to salt-water fish.

remote islands

Any island, by being separated from the mainland, may have differences in the fauna, but it is those whose proximity to the continents are remote in time or space that merit special consideration. There are two classes of remote island, the purely oceanic which has never been connected to other land masses and the isolated continental island which has had no recent contact.

There are many oceanic islands and few are very large. Small islands cannot support as much variety of life as can be found in larger areas, and ocean islands are rarely invaded by any animals that cannot fly or swim. In the absence of competition, birds and other animals develop endemic and sometimes rather unique forms, such as the dodo. The older the island the more of the peculiar species are produced. The more remote the island the fewer the arrivals from the outside world.

Fauna of remote continental islands also are affected by size of land area, distance from other land, and length of time since being connected to the mainland. Thus the largest and longest isolated of the continental islands are the most distinct. There are two outstanding examples, Madagascar and New Zealand. The animals of Madagascar show considerable difference from Africa particularly in including rather archaic placental mammals. This island is famous for its lemurs which, like its civets and tenrecs, have radiated into many forms resembling other animals, just as have the marsupials in Australia. Gaps and variants characterize other animal groups. It is quite possible that the majority of the Madagascar fauna has arrived there since it became an island. New Zealand has an even more unique fauna. There are some very archiac survivals such as a large lizard-like animal long extinct elsewhere. Dominant among its animals were large flightless birds of surprising variety but vulnerable to man, the hunter.

Aquatic Life

Relative to land biota, aquatic life has a fairly simple distribution. The oceans are continuous and freshwater is profusely divided. Temperature and depth within the oceans are reflected by biotic contrasts, however, so that a variety of aquatic regions can be identified. The fresh waters have their relationships primarily with the land biomes. The oceans can be separated into the littoral zone and the open ocean (which has a further vertical dimension) or into tropical and polar zones (Map 3-14). Three combinations will be differentiated and briefly described.

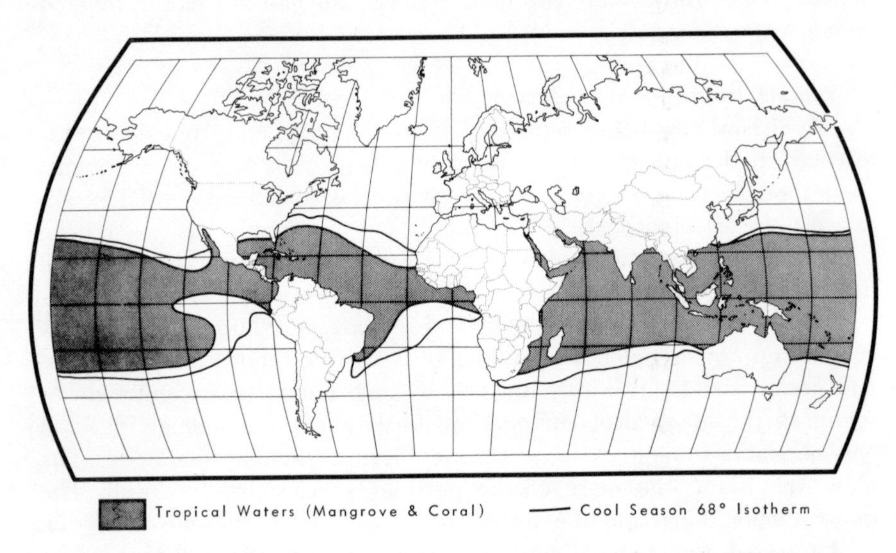

MAP 3.14. Oceanic zones. The interconnections and separations can readily be seen.

Tropical Littoral

The tropical coasts are rich in life, both plant and animal. Coastal mangrove represents an overlap between land and sea. Coral, on the other hand, living in the water, nevertheless is a land builder. The variety of life in the reef seems infinite, involving fish, mollusks, echinoderms, and sea weeds particularly but many other groups as well. Conversely, species populations tend to be rather small so that instead of dominance there is diversity. Shore birds and mammals such as the dugong and manatee are generally related to tropical coasts but also range into fresh waters. Tropical coasts are separated into many units, particularly those on each side of the Atlantic and the Pacific. The western Pacific is more or less continuous with the Indian Ocean resulting in much the largest single province. The many isolated islands have the same sorts of relationships to continental coasts for their littoral life as applies to their terrestrial biota.

Open Ocean

The vast ocean basins are rather sparsely populated but do involve some spectacular life forms. Game fish such as marlin and tuna alongside some giants such as shark, squid, and whale are common. There are sea birds such as the albatross which spends most of its life in the open ocean. Others, such as the petrel and the gannets are among the world's more numerous birds. Floating sea weed, such as is found in the Sargasso Sea, is of rather limited occurrence. Inasmuch as the oceans are continuous, the divisions of the open ocean into basins is only a matter of degree, especially affecting those forms which tend to avoid the temperatures below that experienced at the Cape of Good Hope.

Polar Waters

In contrast to the land, the colder waters have particularly abundant resident populations due to higher solubility of oxygen and important minerals. The variety of species is restricted, but among the smaller forms the numbers are enormous. At the base of the food chain is the submicroscopic plankton. Small fish such as anchovies and sardines run in vast schools. Larger fish include cod and salmon. The top of the food chain is occupied by marine mammals such as seal, walrus, and whale. Use of the coast or of rivers for breeding sites by many predators (seals, birds, and salmon) is most notable. There are littoral life forms along the cooler coasts as well as in the tropics, including oysters and clams, sea weed and salt marsh. The polar waters can readily be divided into Arctic and Antarctic provinces with the former further slightly differentiated into Atlantic and Pacific regions. In the north, salmon and walrus are characteristic while the penguin is exclusive to austral waters.

Because more than two-thirds of the earth's surface is ocean, the volume and variety of aquatic life undoubtedly surpasses that of the land. Man is a land creature; therefore he knows more about the lands and these areas are more important to him than the sea is. There is an increasing interest in the sea, however, and in the future it will require more attention than has been the case so far.

Chapter **4**

Factors in Biotic Distribution

Origin of Biotic Regions

Several generalizations can be made concerning the variation from place to place of the biotic assemblages and how this variation came about. Briefly involved are two kinds of isolation, landmass and climate, opposed to which are the effects of connecting, corridors. The assumption is made that, given access to suitable conditions, plants and animals will migrate into an area, subject only to short run mechanics of dispersal, and there seems to be ample evidence to support this assumption. The limits, then, of a biota must be in the form of a barrier composed of unfavorable conditions. The division between land and water is one fundamental isolating mechanism. Climatic contrasts with the related soil conditions represents the other isolating factor. It must be recognized that there is probably no such thing as an absolute barrier; living things sporadically do cross immense distances as the life on isolated islands proves, a fact discussed in the previous chapter concerning species distribution. But whole biotic assemblages do not jump together in this way so that effective barriers to biota are quite real. However, distant regions are often united to some degree by some kind of corridor. Such corridors on land are generally formed by mountain chains along which ribbons of favorable climatic conditions afford passage. The marine equivalent is a strait.

isolation of landmasses

Isolation of a landmass means a water barrier of more than a temporary nature. The result is that a substantial part of the biota on either side of the barrier will not span it even though some elements do. Obviously, because there are no absolute barriers, there will be degrees of separation related to the duration and size of the gap. Those organisms most effective at dispersal over long distances, such as birds, bats, ferns, cattails, and many weeds, will be best and most frequently represented on either side of water divisions. Several major instances of the isolation of landmasses have been suggested as being reflected in the distribution of land plants and animals with respect to whole continents and even hemispheres. The more ancient divisions persist in plant patterns because of their slower dispersal, while more recent isolations seem to affect animal patterns more strongly.

Many lines of evidence point to the separation during earlier geological times of a southern landmass called Gondwannaland from the more northern land areas. Essentially this means that North America and Eurasia, with the exception of India, were not in contact with South America, Antarctica, Australia, and India until sometime in the mesozoic era. This is possibly reflected in the contrasts between the holarctic flora and that of the rest of the world. Particularly the conifers, which are ancient and were well-differentiated while Gondwannaland was still isolated, suggest such a separation. There is a clear division between the holarctic families, Pinaceae and Taxaceae on the one hand and the southern families, Araucariaceae and Podocarpaceae on the other. Two other families, Cupressaceae and the relict Taxodiaceae, are divided at the generic level. That this is not primarily a climatic relationship is indicated by the fact that podocarps and southern Cupressaceae, in particular, range from Antarctic tundra to tropical rainforest. In fact, the holarctic flora, in general, is quite distinct as has been described earlier. Nevertheless, certain very old relationships from north to south are present including the northern *Fagus* and southern *Nothofagus* pair, *Libocedrus* in the south which is most closely related to *Thuja* and *Calocedrus* of the north, as well as *Austrotaxus,* a member of Taxaceae in New Caledonia, and *Athrotaxis,* a Taxodiaceae in Tasmania. Furthermore, many angiosperm families are virtually worldwide with most holarctic genera having related genera within the tropics. Apparently the north-south isolation has been far from complete with respect to plants in general, and much of the observed contrasts can be explained in terms of climatic factors to be discussed below. Nevertheless, there may possibly remain an element of distinction inherited from the time when Gondwannaland was separate from the northern landmasses.

Manifestly, the fragments of Gondwannaland are today separated by whole oceans. Particularly are South America, Africa, and Australia mutually isolated. Antarctica is also isolated but essentially lacks terrestrial life. Only through thousands of miles of the holarctic can any conceivable land connection be made between the inhabited southern landmasses. A change of climate could bring the Old-World tropical regions into more intimate contact but no such possibility exists for South America. Whole families of plants such as Cactaceae or Dipterocarpaceae have developed exclusively within a single realm with only minor subsequent leakage externally. Interestingly, the Antarctic flora as a complex does span the oceans (only a few elements in Africa). It appears that Antarctica had a role in this migration which is strictly intrusive into the tropics of South America and Africa but not Austromalesia. Perhaps the fragments of Godwannaland were only more recently isolated in the higher southern latitudes.

Australia has been for a long time and still is isolated from other landmasses. The arid flora there is indeed unique, but this can be explained on climatic grounds, inasmuch as the humid flora of Australia is not unique. It is the animals that set Australia and the nearby islands apart. The mammals there are completely different from those of neighboring Asia, being entirely marsupial or more primitive, with the exception of a few clearly intrusive placentals. The marsupials have developed forms for a wide variety of niches paralleling the placental mammal adaptations elsewhere. Other animals are also distinct. Today the water gaps separating the Australian region from the rest of the world are small but involve an archipelago of islands. Asiatic animals dominate the land as far as Java, Borneo, and the Philippines, areas which are connected by shallow water which has recently been dry land. Wallace's Line marks the end of this dominance. Australian animals dominate in New Guinea, Halmahera, and the nearby Moluccas. A tissue of shallow water is also present here. Where the Australian fauna terminates is called Weber's Line. The area in between, Celebes, Timor, and Flores (the Lesser Sunda Islands), has sporadic representatives of both faunas but not much of either plus some unique survivals such as Komodo dragon, the world's largest lizard.

South America has a similar faunal uniqueness together with important elements shared with North America. A number of peculiar mammals are native to South America such as the armadillo and the sloth. These are placentals, but South America also has a significant marsupial population, the only such group outside of Australia. There are notable distinctions among the fresh-water fish as well. Recent fossils include many additional peculiar mammals showing that in isolation there evolved a range of forms to fill the different ecological niches available. Inasmuch as placentals dominate the unique mammals of South America, the isolation apparently had its inception later than that of Australia. Recently, furthermore, the two Americas have resumed their land contact with particular impact on the South American fauna. Cats, elephants, deer, horses, camels, and other widespread northern types invaded southward, followed by the extinction of many endemic animals. There was an exchange, with a possum and a sloth ranging widely to the north followed more recently by the expansion of the armadillo. The development of a land bridge, naturally, means that for marine life there is now a barrier (which may be destroyed by the construction of a sea level canal, with unknown consequences).

There is no present-day land connection between Eurasia and North America. There has been a union in the recent past as evidenced by the similarities of the flora and fauna. That there is a small degree of difference in plants, particularly, is only natural in view of the current separation.

climatic isolation

The isolating effect of climate can be as profound as that of the separation of land and water bodies. In the world there are major contrasts in available heat and moisture. Other factors such as humidity, cloudiness, wind, and pressure (elevation) have effects but these are minor and doubtfully ever act independently as real barriers. Either extreme of heat or of moisture; wet, dry, hot, or cold can act as an obstacle to the penetration of differently adapted biota. Just as important, each of these, with the exception probably of warmth, can also exist, thanks largely to surface configuration, in the form of a bridge. On the other hand, terrain might introduce a climatic barrier just as effective as the great climatic zones.

Climatic Frontiers

Climate is a continuous variable so that partial or complete biotic discontinuities related to climate require further explanation. Individual species have finite limits but, through speciation, taxonomic groups and even biota can extend across gradients uniting extremes of variation. There are desert oaks and forest oaks, even rainforest oaks. The Australian flora extends from snowy peaks in the south to hot tropical reaches in the north. Nevertheless, biotic frontiers do exist and there seems to be two ways that this comes about.

One of the ways in which a biotic frontier may be produced results from a former separation. Development in separate and distinct environments tends to produce different abilities and tolerances. When these separate biota manage to impinge on one another, each faces both an increasingly different environment and an established population eminently adapted to that environment. Invasion requires both effective adaptation and successful competition. Obviously, a certain amount of delay in mixing is likely here with therefore the persistence of a sharp boundary zone. Certain elements will be better suited for penetration of the new territory than others, of course. The sharpness of the oak-pine boundary with the tropical legumes and cactus found all around the mountains in Mexico and Central America may well reflect the former separation of North and South America.

The other origin of a biotic frontier concerns the effects of extremes and their interplay. Within any continuum, innovation (speciation) in any ubiquitous element may relate to one of the extremes, which could result in a division in the continuum. Division would be most common near the median because, taking temperature as an example, no one organism can at once be especially adapted both for heat and for cold. A new type adapted for cold would be better suited for cold survival than the antecedent form which was either unspecialized or better suited for heat. But yet another type appearing between these two in adaptability must therefore be less well adapted to either extreme and would tend to be superfluous. The accumulation of such divisions could then produce an increasingly sharp frontier separating the two extremes of the continuum, particularly if any sort of mutual interdependence is developed between the elements on either side of the frontier. That such a phenomenon has occurred is indicated by the biotic separation of every landmass region with respect to two or more extremes as has been described previously.

Climatic Barriers

If there are discrete biotic assemblages related to climatic factors, the possibility exists that there should be climatic barriers to their dispersal. Either temperature extremes or moisture extremes might become insurmountable obstacles to a whole biota as a complex and a number of major examples exist, in addition to many minor local effects (Map 4-1).

Cold is generally a barrier to tropical life forms. All of the floras of the tropical provinces, both humid and arid are firmly excluded from the higher latitudes with only sporadic elements breaking through. Plants and animals of the tropical littorals are similarly held in check. Less sharp, perhaps, are the limits for tropical land animals and yet even for these an important difference exists. The palearctic fauna differs considerably from the palaeotropical fauna despite their open contact with one another. Conversely, the cold floras and faunas, particularly the terrestrial flora and the marine fauna, remain largely outside of the tropics and are thus differentiated into boreal and austral provinces.

Arid and humid can only apply to terrestrial life. The one great arid barrier is the Sahara which extends unbroken from the Atlantic to the

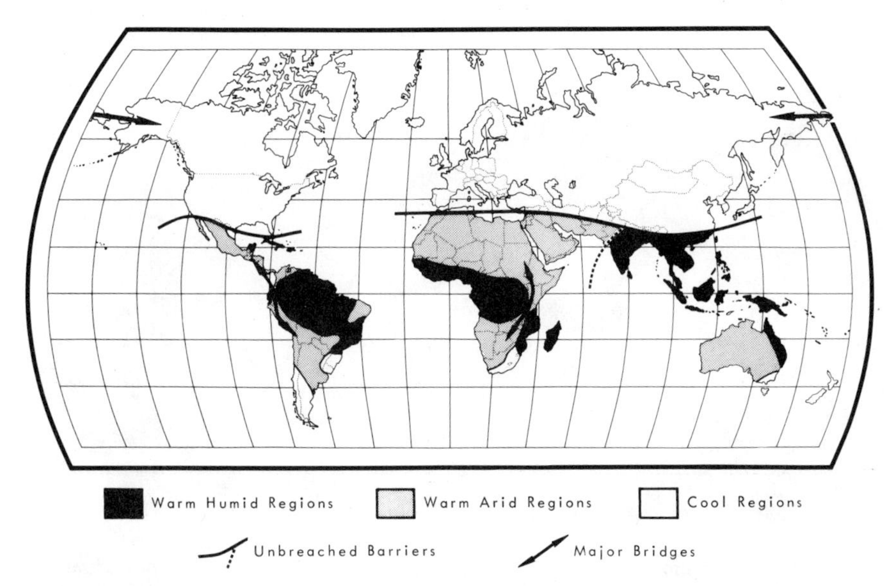

MAP 4.1. Major climatic barriers and bridges for the spread of biota. There is no passage for warm biota across the cool regions nor for humid biota between Africa and Asia, as well as for arid biota from Australia to other regions.

Indian Ocean. Animal variety is greatly restricted in extreme environments and relatively few have crossed between Africa and Eurasia (Map 4-2). Even more, this arid obstacle has isolated the humid flora of Africa. Humidity, on the other hand, separates the dry flora of Australia from Asia. That the water barrier here which has divided the animals is not a plant barrier is seen by the broad plant similarities between southeast Asia and New Guinea. Similar humid zones in Africa and America have had only a partial effect and cannot be considered true barriers. The arid floras of America and Asia are divided by areas which are simultaneously more humid and colder.

Local climatic interruptions are many, if less absolute than the great climatic zones referred to so far. The arid zone between California and Arizona, for example, restricts communication between life in the humid areas on either side (Map 4-3). Conversely, the Sonoran arid flora has limited contact with the arid zones of interior Mexico. For littoral biota the presence of fresh water at the mouths of major rivers represents a major climatic barrier. Even a tongue of cold or warm water, such as the Gulf stream, can divide the population on either side.

Considering, then, the great landmass units with respect to the pattern of water together with the climatic divisions, it is easy to account for the various distinct biotic provinces which have been earlier de-

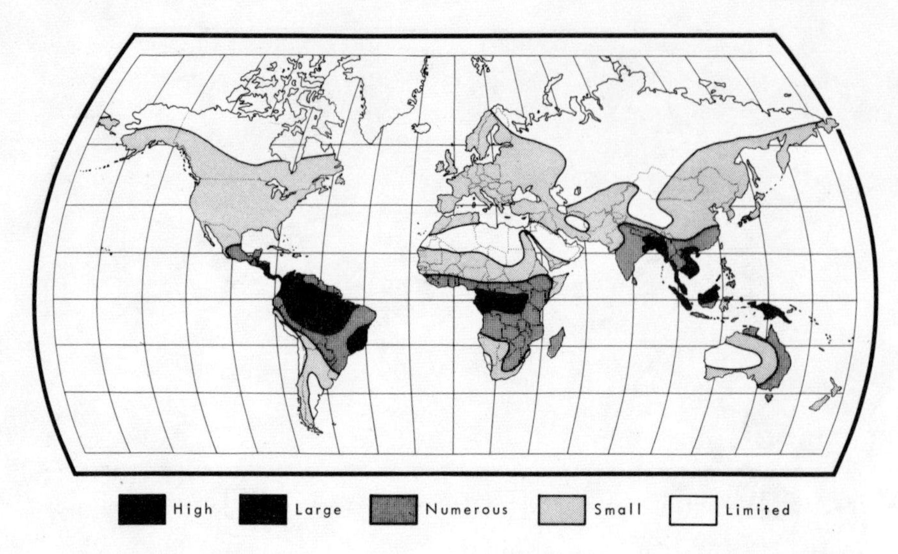

MAP 4.2. Animal variety in very generalized distribution. Individual animal groups vary somewhat from the overall picture.

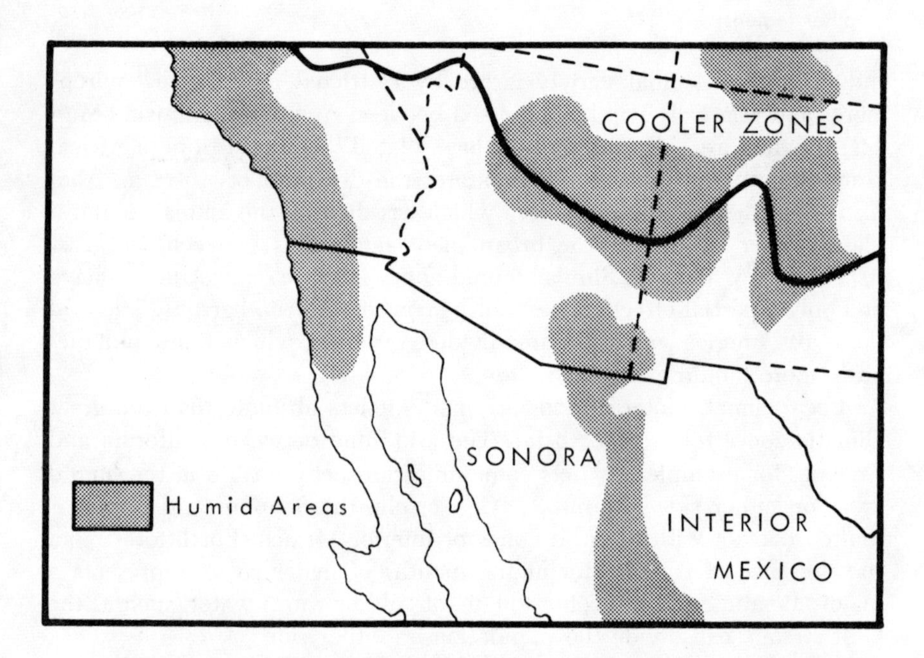

MAP 4.3. Local barriers in western North America. The humid highlands of California and Arizona are completely separated while Sonora communicates only slightly with the interior of Mexico.

scribed. There are the three humid tropical regions centering respectively on the Amazon, the Congo, and Indonesia. There are three arid tropical regions: America, Africa, and Australia. There are two humid extra-tropical zones which, with the omission of the boreal subrealm, must be partly divided among different continents. Finally, there are two arid middle-latitude regions.

Corridors

In various ways, largely separated biotic regions can intercommunicate through accidents in the configuration of land areas. Such connections or corridors may involve one or more of several factors, four of which can be readily identified (Map 4-4).

Mountain ranges offer, as a result of their characteristic variety of local climates, avenues for migration across alien climatic regions. Higher elevations are cooler and can therefore provide a bridge for cool area biota across the tropics. It is a long way from one side of the tropics to the other, however, and on each side there are already resident populations, so tropical mountain corridors involve only a selective interchange of extra-tropical forms characterized by progressively attenuated peninsulas equatorward from the middle latitudes. Mountains also harbor dry zones as a result of local rainshadow and these places have effectively united northern and southern hemisphere arid biota in both America and Africa. The width of the humid tropics being much less

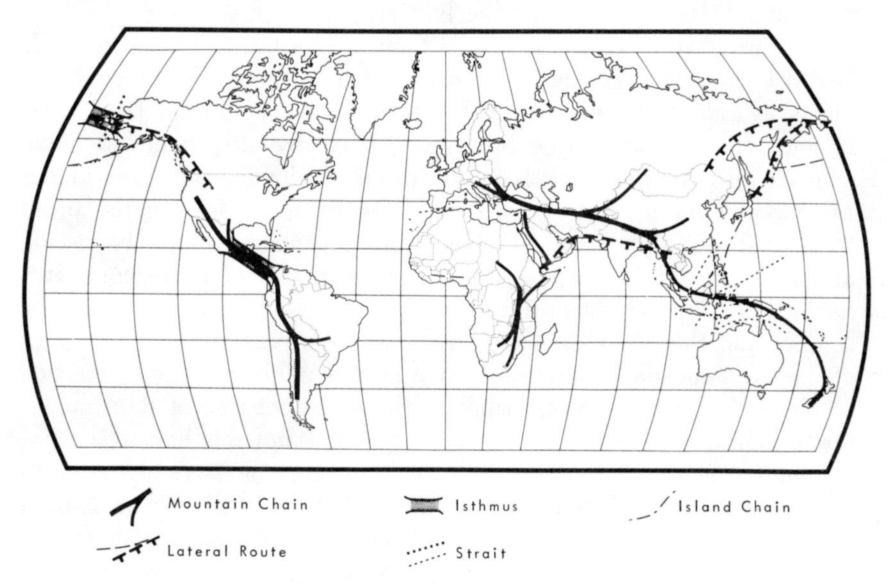

MAP 4.4. Climatic corridors.

than the tropics in general, probably a majority of the biotic elements from either hemisphere has become established on both sides. But mountains include wetter slopes from orographic effects so they may even form a humid bridge across an arid expanse. The greatest humid link along a mountain chain is between Europe and China. The continuity of humidity along the highlands bordering the Red Sea across the Sahara is tenuous at best and only a minimum of elements have travelled this route. Other more local examples of mountain corridors of similar types can easily be discovered with a closer inspection of the map.

Lateral corridors occur where two similar biotic regions both border on the same contrasting region but are themselves physically separated. Essentially this refers to coastal corridors, one example of which is the South Asian corridor. There are important similarities between the animals of the oriental realm and of the palaeotropics on either side of the Indian Ocean. Apparently migration has occurred between these two regions along the southern edge of Asia. Similarly, the littoral biota of East Africa is united with the East Indies along the same coast. Another major corridor lies around the northern side of the Pacific Ocean. Milder oceanic temperatures along the coast have helped allow extensive migration between the Old- and New-World middle latitudes. Fossils in Antarctica suggest that the Antarctic coast has in the past been a route from the western to the eastern Pacific. Again, lesser examples can be advanced, as between Chinese forest lands and Korea or the more arid biota of Iberia and the Balkans.

A third sort of corridor is an isthmus. The most obvious example lies in Central America linking North and South America where simultaneously mountains have provided avenues for migration from more remote regions. The Bering Straits area has obviously operated as an isthmus for extensive periods in the recent geologic past. Any isthmus can have a uniting function. At the same time, considering the possibility of dispersal over short distances of water, many straits and island arcs have played a transportation role akin to that of an isthmus. Particularly is this true with respect to plants through the East Indies. The Antilles and the Aleutians have also hosted the passage of a few forms of life. At some time there must have been a more intimate connection between New Zealand and islands to the north because of the similarities which are found. That in recent geologic times this has not been a continuous land connection is shown by the progressive decrease in land fauna eastward from New Guinea, but that this has not been strictly dispersal over water distances as now constituted is shown by the endemic flora of New Caledonia, today no more removed from its neighbors than they are from each other, while beyond in Fiji there are many plant species identical with those of New Guinea or even Southeast Asia.

Essentially similar to an isthmus but involving water is a strait. Perhaps the greatest example today is the Bering Strait uniting the Pacific with the Arctic Ocean. Between Australia and Asia there are several alternative passages but the tropical waters of the Pacific and the Indian Ocean are only connected through narrow water channels. Many smaller straits such as the gap at Gibralter or at the mouth of the Baltic Sea connect inland arms of the ocean with open water.

With an understanding of the various isolating and uniting factors can come an understanding of the distribution and relationships between the various biotic assemblages. There are recognizable groupings of taxonomic entities and these have migrated at least in part to distant places. They have exchanged elements with their neighbors and, more significantly, they have exchanged elements with more remote but environmentally similar biota. Simultaneously, there are some taxonomic elements with a cosmopolitan distribution masking somewhat the provincial distinctions which have been illuminated here. Any living thing whose individuals or whose propagules can be dispersed over great distances through the air or in the water may supercede the limits considered above. Some bats, birds, and insects fly on journeys of thousands of miles. Seeds or spores of cattails and ferns drift with dust. Strand plants and small hardy animals such as rats and lizards can survive rafting to distant reaches of the ocean. Some tropical plants, including banyan trees and the mimosa family, seem to have no distance barriers. More than any other species, man has found his way to far corners of the earth. Along with man then have gone his intimate associates, mostly pests such as flies, fleas, roaches, dandelions, plantains, and sour clover. With man have also gone dogs and cats and a host of domesticates. But in nature there have also been sporadic long distance introductions which, taken separately, bedevil any logical explanation of pattern but which do not negate the predominant rationality of the geography of life.

Implication of Biotic Patterns

Aside from characterizing area, biotic patterns are worth studying because of the insights they yield about a variety of phenomena. To a considerable extent, the implication of biotic patterns overlap with what can be learned from the distributions of individual populations, an expected relationship inasmuch as a biota is an aggregate of individual populations. Thus additional understanding is given to environmental change, variation in climate, and surface configuration, together with continental migration. Likewise, more light is shed upon changes in life forms, the role of environment in evolution.

Environmental Change

Climatic change is more clearly revealed by whole biotic assemblages than by individual population patterns discussed above. Individual populations might come to transcend the limits of a biotic assemblage or lag behind when adjustments are occurring but the biotic complex tends to maintain its identity against neighboring groups. The presence of currently isolated plant and animal complexes from Utah through Arizona to Chihuahua in mountain areas, for example, is highly suggestive of climatic change indicating that at one time suitable conditions existed at lower elevations for a continuous expanse of this biota across the area in question. If only sporadic species were involved in this discontinuity, it might easily be explained as sporadic invasions.

There is much evidence, fossil as well as geographic, that climates have fluctuated repeatedly, both in terms of temperature and of moisture in the last few million of years, with the consequent dispersal of biota. Such changes have undoubtedly facilitated the exchanges across the equator of arid biota. Clearly, middle-latitude forests have migrated to latitudes high enough for their continuous interconnection across northern land masses. Even some holarctic arid plants may have thus been enabled to make a lateral trip around the North Pacific. Perhaps many European species got trapped and perished along the coasts of the Mediterranean along with a southern migration, but these movements generally aided dispersal of biota to distant, suitable and now isolated environments. These changes in climate are associated with the well-known glacial epochs of the higher latitudes.

Changes in landform, slow in relation to evolution of life, have had some discernible effect on present patterns which conversely help to suggest where such landform changes have recently occurred. An example can be found in southern South America between southern Chile and southern Brazil. These two areas, separated today by a broad arid band, nevertheless share related or even identical forest species. The kinds of climatic shifts visualized for the recent geologic past would not, in terms of present terrain, bring these two life assemblages together. Rather, if the mountains, which are responsible through rainshadow effects for the intervening dry zone, were not present, contact would readily have occurred. These mountains, from other evidence, appear to be geologically quite young so the evidence is in agreement. Also, it is thought that the Cascade Mountains are quite recent and that the western American forests formerly extended far eastward.

Continental Drift

Biotic patterns have often been cited in connection with the movements of the major land masses since some time in the mesozoic era. The rapidity with which most animals can migrate when corridors for travel exist limits the discussion largely to plant patterns. It is utterly doubtful whether the distribution of plants can yield any conclusive evidence in favor of continental drift but the reverse of this may well be true; continental drift may help explain the distribution of plants.

There are many plant patterns that seem to relate to the areal units of continental drift. The holarctic flora is sharply differentiated from the remaining floristic regions which, though now in broad contact with the areas occupied by the holarctic, correspond to the elements of Gondwannaland. At the same time, far reaching similarities exist between the humid tropical floras of the three widely separated parts of the former southern continent and the same may be said of their arid floras and their temperate floras. There is palaeobotanical evidence that these now separated groups have in general been in existence through at least a substantial part of the geological time that has elapsed since the breakup of Gondwannaland.

Against these arguments are several considerations. In the first place, the great majority of plant families are, in fact, of worldwide distribution. Unless such families have been in existence since before the breakup of Gondwannaland, a rather effective dispersal must be visualized. Besides this consideration there is also the implication of the general division of regional floras into contrasting parts. Intertwined in the same landmass are distinct arid and humid floras in all three parts of the tropics, and there is a sharply differentiated cool-climate flora of the southern hemisphere that has intimate relationships with the Malesian tropical highlands. By analogy, the contrast between the holarctic and the adjacent tropics then may be the result of similar forces. There are alternate explanations to continental drift available to explain the floristic patterns of the world.

The truth of the matter probably is that all of the above considerations play a part in accounting for the distribution of flora. Problems of adaptation and competition may well have kept substantial floristic elements within their ancient homelands particularly those elements least effective at long-range dispersal. On the other hand, where environmental adaptation was not a problem, as between two humid tropical regions, the greatest separations may have been sooner or later conquered, particularly by those elements which are most efficient in their dispersal. Some groups of plants are undoubtedly very old and have ridden with the continents as these have shifted position. Other plant groups are, of course, regionally unique.

Evolution

Biotic patterns illuminate the process of differentiation through parallel development. Each separate biota in a particular sort of environment necessarily confronts similar opportunities for life. A study of any pair of analogous biota may reveal similar life forms derived from dissimilar origins. Many remarkable examples exist of this parallel development of biota, more remarkable as an assemblage than the individual similarities might be in isolation. Well-known is the development in Australia of marsupial analogs of a whole series of placental forms. These are bear, wolf, rat, mole, and even the kangaroo, for all of its unique appearance, lives much like, for instance, an antelope. Another set of parallel developments of animals in South America has now ceased as the animals are largely extinct. Between the arid lands of Africa and America many similar but unrelated plants and animal pairings can be made. American cactus is mimicked by the euphorbias; agave resembles the aloes; kangaroo rats look like jerboas; and even the antelope of Africa are more related to cows than to the American pronghorn. Various needle-leaved but flowering plants of Australia resemble pine in general growth form while certain small-leaved evergreen beeches of Chile and New Zealand have essentially the same aspect as hemlock. Each of these is a part of an assemblage which also resembles the corresponding biota. In some cases of course there are also included in two separate regions elements which are closely related. Furthermore it must not be supposed that all forms are paralleled, only that remarkable similarities have arisen in response to similar opportunities.

Regional Character

The mere characterization of a region may not solve any problems but is, nevertheless, a paramount function of its biotic description. The sum total of what lives in each region is finite and is not at all the same as that of any other region. Essentially, the life of an area is what gives it its personality. Australia is kangaroos, koalas, eucalyptus, and acacia, to mention some obvious and well-known elements. Oak, pine, maple, elm, deer, wolves, wildcats, and rabbits together spell the holarctic, familiar to its residents, exotic elsewhere. The inventory of a region supplies its potential resources, what is there, what is not there. The least that can be known about life in an area is the content of its biota. Surprising as it may seem, for much of the world this is still imperfectly known.

Chapter 5

Life Communities

In any particular area at any particular time there exists together a particular set of species which, taken together, can be called a life community. Essentially the community is a subdivision of the biota, but it is more; *the community is a specific grouping while the biota is a general assemblage.* Because the community is specific, it is necessarily more local in area and there are a great many of them. Although the community concept is widely used, there is strong disagreement as to what it means so that the student who pursues the subject should be prepared for controversy. Only a geographically oriented consideration of the subject will be given here, the meaning of the concept, the relation of community to environment, and the geographic use of the idea of community.

The Nature of Communities

There can be no question that a rather limited set of species coexist in each local space; the idea of community is that such sets are not random relationships. Thinking varies from the proposition that the relationships are obligate, as in an organism, to the position that the groupings are in fact random. It is probably safe to say that neither extreme view is correct and that there is some intermediate condition. Relationships exist and elaborate studies have been conducted in order to establish their nature. This has been followed by the designation of a variety of community types.

Species Relationships

The areas occupied by various living populations often overlap. There are many degrees of coincidence and ways that this may come about. The ecologist studies these relationships in elaborate detail; for the geographer there is an important spatial aspect of this larger ecological problem. How much of the overlap is random and what are the effects of discontinuities?

The overlap of life territories can be readily described. The two may coincide or one may be entirely included within the range of another. More commonly both extend beyond the range of the other to some degree. For a community, it would be expected that a high degree of correlation exists among the members' areas. Terms have been applied to the degrees of relationship such as *exclusive* when a species is nearly limited to one community, *selective* if the species has a majority of its occurrence in a particular community, or *preferential* if the species is only more frequently present in one of several communities. Aggregates, it has frequently been shown, are found where several or many species are selectively or even exclusively present in the same area. With these species may be others whose presence has a high level of *constancy* but which also occurs abundantly elsewhere. It must not be supposed however that everywhere or always do life forms correspond to those distributional relationships because many areal relationships are certainly more or less random.

If communities exist, then, there must be a reason, something must be binding the various elements together. In some cases there is a symbiotic relationship such as that between yucca and a particular moth that polenates it. Neither can live without the other. Many delicate shadings of symbiosis exist but probably this idea should not be extended to non-specific relationships such as the need for shade, which can, nevertheless, be an environmental requirement. More often than symbiosis, the binding force may be a common need, such as that for heat, moisture, soil acidity. That is, the predominant community-forming factor is parallel response to the environment which in turn varies from place to place.

Independent species each respond to the environment differently than do any other species so it may be argued that, unless there is an obligate symbiotic relationship, their ranges will be independent, that is, randomly related. Indeed, such appears often to be the case. Much of the apparent correlation seen results from discontinuity. The limits of many species will coincide where a sharp environmental shift occurs throwing these species together. But the environment is not divided everywhere nor in a consistent way by discontinuities, so that the reality of many communities degenerates into the fortuitous existence of local environmental contrasts. Where, locally, swamps may be sharply marked off against the uplands, over a larger area there may be variations in the degree of swampiness and the sharpness of the swamp edges. Such considerations weaken the concept of community, but there is another factor that supports the concept.

Although environment is dominantly a continuous variable, there are several couplets of environmental poles. Each environmental element varies between opposed limits. The response of life to such polarization has already been discussed with respect to major floristic divisions. There is reason to believe that a fairly limited set of poles, such as wet-dry, hot-cold, acid-alkaline, sunny-shade, can be recognized and that life forms tend to divide between such poles. There may also be certain critical limits such as frost-no frost, drought-no drought that divide the life content of space. Each flora therefore is locally divided into communities corresponding to the contrasts present within its territory. Where discontinuities exist, they may sharpen these divisions remarkably.

Types of Communities

The extent and complexity of community grouping can obviously vary and this has led to the development of nomenclatural distinctions. Further taxonomic names have been applied to each community by some workers. Others treat the groupings less formally by applying individual names, thus affirming at least that there is some kind of relationship. A few of the more important designations will be considered.

The basic unit of the community is the *association,* a term usually applied to the plants alone but not necessarily so. The association is characterized by particular dominant and index species, each association therefore being unique. One may speak, for example, of beech-maple-hemlock forest alongside pitch pine-scrub oak or elm-red maple associations. A variety of small plants and animals will be common to each of these tree groupings. It is possible to consider distinct developmental communities as associations (or *associes*) such as a hawthorne-cherry scrub or a cedar "brake." According to some students there may be hundreds of associations in an area such as France.

There are several ways to subdivide associations. Within the larger structure one can discover many niches such as the canopy, the understory, or even the soil as well as smaller habitats such as bark, rock surfaces, and local variations in the soil configuration and composition. (There is almost no end to the possibilities of microenvironments which can even include the skin or alimentary canal of a particular species of animal, but geographically one's attention would be limited to identifiable space.) Another division of the association is in terms of variants or *faciations* of the larger combination. One may recognize the integrity of a particular association but nevertheless may discover that its content is variable. Beech, maple, hemlock, and basswood occur together widely but there are extensive areas where beech is absent, others where hemlock is absent (Map 5-1). The term northern hardwoods is sometimes applied to this association thus including some less common but widespread constituents. Alternately each of the faciations may be considered a separate association. Disturbed communities may also vary in content from the undisturbed stand and may be considered as aspects of an association.

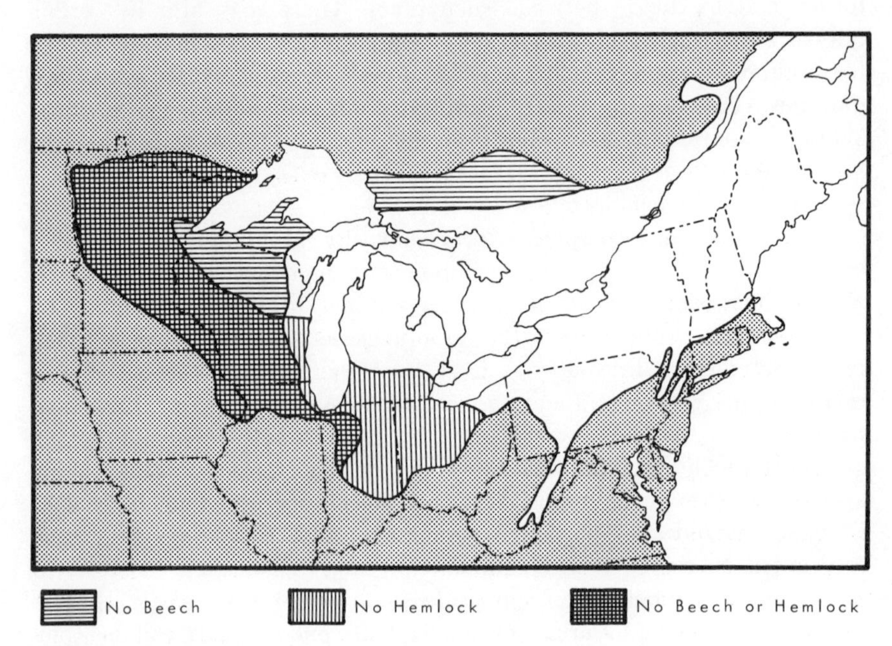

No Beech | No Hemlock | No Beech or Hemlock

MAP 5.1. Faciations of the beech-maple-hemlock-basswood association. It it significant that a majority of the indicated region is an ecotone (transition) to other dominants (spruce-fir; oak-hickory; southern hardwoods) so that it may be considered as a faciation of a larger entity.

Aggregations of associations may be made. The most inclusive term applied is *biome* which particularly includes animals but also groups together developmental stages, faciations, and local habitat variations. The biome is in fact essentially synonymous with the biotic province as it has been treated above. The idea of *climax* has been applied here also with reference to the ultimate stabilized association after all edaphic and disturbance factors have been overcome. The climax can also be taken to include these related combinations and can be used as broadly as the biome or in a more restricted areal sense. The Braun-Blanquet school works with hierarchies of associations through *alliances* and *orders* to *classes* that are essentially confined to a particular habitat type Thus a community class might correspond to the swamps of eastern North America. Implicit in much of this nomenclature is the assumption that community and form correspond, which may not be strictly true, particularly for the larger groupings.

Factors in the Origin of Communities

Of the many thousands of species that exist on this earth only a few assemble in each locality. The reasons why this is true are many but can be grouped under two headings: external influences and dynamics of the biota, that is, stresses and opportunities which impinge upon life forms from without and the interplay between these life forms based on their own intrinsic abilities. What is being discussed here is ecology taken with an eye to the spatial dimension.

the environment

Life as man knows it lives at the interface between the surface of the earth and the atmosphere. Sometimes this life zone is called the biosphere and it obviously has strong geographical qualities. Attention is therefore immediately directed to either side of the sandwich which encloses life as being the primary subdivisions of the environment. The term *edaphic* refers to the solid foundation or soil on or in which life is based. Ocean and lake bottoms can be appended here to the extent that they are an influence. Conditions in the fluid mantle, particularly the air, are described by the term *climate.* Each of these environmental aspects vary importantly from place to place creating differences to which life responds, in part, by dividing into communities based upon tolerances or requirements for particular conditions. Interestingly, wherever there is a lateral boundary or edge of any sort, this too becomes a special environment.

The Role of Climate

Given any climatic gradient with distinct extremes, there will be different plants and animals found grouped at either pole (See Map 3-1). Climatic factors have been discussed earlier where it was stated that temperature and moisture effects are dominant with all other aspects of climate such as humidity and cloudiness often linked. Just as population patterns can be profitably analyzed in terms of climatic parameters, so can community patterns. The division of communities with respect to climatic gradients is affected by the steepness and regularity of the gradient.

Not only are climatic variables continuous but they are generally characterized by gentle gradients with only sporadic irregularities. Over vast distances there is frequently an irregular progression of populations with no obvious hiatus to separate the whole into communities. Starting for example with southern forest trees and proceeding north the sequence is approximately to lose first laurel oak and then add sugar maple, lose loblolly pine and then after a distance lose southern red oak. The addition of basswood is shortly followed by the loss of sweet gum and holly and the addition of hemlock. One by one the more northern species appear, yellow birch, fir, and then spruce, while tulip tree, hickory, and black oak drop out. Paper birch is the last important species to appear while well to the north the so-called northern hardwoods disappear one by one and the last non-"boreal" species to drop out is white pine. The boundaries of these populations are not parallel so the sequence varies from east to west. There are some groupings of limits to be sure, but overlap is the rule. Many similar climatic transects have been made and these tend to indicate that sharp community boundaries are not usual in response to climate.

Climatic gradients are not always gradual. In mountainous terrain many abrupt changes occur as between north and south facing slopes. Very sharp community boundaries are not surprisingly found in such places. For various reasons strong climatic contrasts are found across even level areas. Where southerly winds bring moisture north from the Gulf of Mexico, a distinct climatic boundary extends poleward through Texas from the west side of the Gulf. The northeast side of the "prairie wedge" locally coincides with the temperature gradient across Minnesota and Wisconsin and promotes community contrasts there. But these sharp gradients are only sporadic and so the conclusion emerges that climate is not a very effective differentiator of life communities, even though climate-related variations are easily demonstrated. The same must be said of the climate of water.

The Role of Soil

Almost an infinite variety of soil conditions can be described from place to place. There are gradual or zonal characteristics which are strongly related to climate but, more important, there are intricate local variations based on drainage and parent material contrasts. These edaphic distinctions are responsible for the differentiation of numerous well-marked communities. The edges of swamps or bodies of sand or granite outcrops or valley alluvium can be razor sharp. Furthermore, any particular soil type may recur time and again, so that the associated life forms are isolated over and over beside other clearly distinguished associations (Map 5-2). Frequently, the edaphic conditions mimic climatic extremes of moisture and, in part, bring related organisms far from their general climatic homeland. The idea of community has its most successful application where local edaphic factors come into play. Similar distinctions in water habitats also apply. Coral reefs locally are marked off from mud flats and sandy bottoms to mention some examples. Furthermore, islands, lakes, and rivers necessarily have communities distinct from their surroundings, the general division of the earth's surface into land and water being a fundamental distinction.

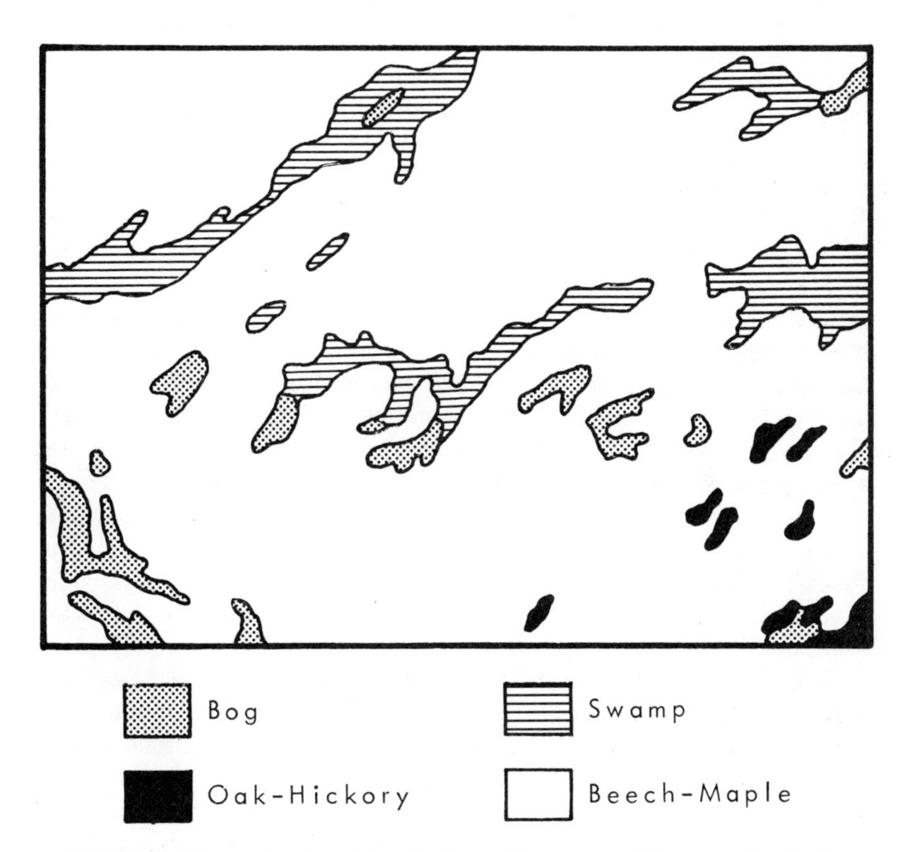

MAP 5.2. An example of local plant-soil patterns. The map is of the original vegetation of a portion of Monroe County, New York (after Shanks).

Edge

Where two different environments meet, there is the possibility of interplay between the two, of which some life forms may take a special advantage. The most obvious edge is that between the land and sea where a great many species of plants and animals have carved out a very particular zone in which to live. Although the tidal zone is only a narrow strip, the length of the world's coastlines are enormous and add up to an important habitat. Any edaphic edge can develop in a similar way and the boundaries between plant formations such as the forest-grassland edge offer comparable possibilities. Disturbance particularly increases the availability of edge, a temporary thing in this case but no less important for it.

the dynamics of life

Distribution of living populations is hardly a static thing, thus an understanding of the composition of any community must take into account the kinds of changes which may occur. One might start with modification caused by some abnormal external effect and proceed to the ways that the resulting community recovers toward an ultimate stability. The first of these can be called disturbance and the second succession.

Disturbance

There are many factors which can disrupt a living community, altering or even eradicating it. These factors include catastrophies such as fire, flood, windstorm, epidemic, and man. More gentle pressures tend to be more or less a part of the normal environment and do not merit the name disturbance. In any case, disturbance causes a change, and many living things not only take advantage of such effects but some may even depend upon them for survival.

Fire is the greatest single instrument of community change. In destroying what is exposed, fire works its greatest harm on plants which gain advantage through increments of growth over the years and thus shifts the balance to herbaceous plants. Also aided are the woody plants that thrive in an open situation. There are even plants such as knobcone pine or some of the cypresses whose seeds require fire to open the cone and affect dispersal. After a fire, certain plants spring up, and into this environment come particular insects and birds such as the quail to take advantage of the special conditions. Thus, there are communities specific to recently burned-out conditions within each biome. With repeated burning the surviving community may be restricted to those highly equipped to withstand fire. The koala is not nimble enough to escape and is facing extinction but the kangaroo has little problem. That fire is a normal part of their environment is demonstrated by the prevailingly advanced adaptation to fire conditions of the Australian plants which fact was mentioned above. Elsewhere, periodic fire has largely eliminated the woody plants expanding the grasslands highly suited to the great herds of grazing animals. These communities are sometimes called *fire climaxes* because of the balance that has obtained between them and endemic fires.

Much damage is caused by floods. What is not drowned may be washed away. The soil is eroded or new soil is deposited so that the floodplain becomes a particular habitat even though floods in many places are of irregular occurrence. Some plants not only survive such treatment but regularly spread their seeds through the floodwaters to germinate in masses when the flood subsides. Stands of cottonwoods and willows thus tell of inundations in the past.

Windstorms take their toll of trees, some of which were already weakened, and gaps are left or the soil is even thrown open. Some species of kauri in the South Pacific or white pine are designed to spring up in such spots and gaining a foothold, they may dominate an area for many years, the isolated ones just a part of the forest scene but groves showing where a violent storm struck. The impact of man perhaps most duplicates the workings of the wind (when his tool is not fire) and the same species increase in his trail.

All of these things and others break up the continuity of the plant cover creating numerous new edges. The edge itself is an important habitat as was noted previously. Plants and animals live in or use the edge in important ways. Bringing leaf growth within reach is a significant function of the edge, promoting the expansion of browse animals. Grazing animals can also take advantage of the forest for cover even though their food is in the open. Deer are a well-known edge animal whose numbers have been vastly increased by man's disturbance of the forest. But disturbance certainly occurred before the advent of man and many life communities have developed in relation to it.

Succession

Given a bare surface or a disturbed area, life will not only occupy the area, but changes will tend to occur. From each initial condition there will be, in effect, a specific sequence of life forms leading to a specific stability or climax, according to theory. Reality may be less specific but changes of this nature do occur and any lack of development is in itself significant. Involved here are the three factors, competition, persistence, and availability, each of which has a role in explaining the life content of an area.

Ongoing succession works through competition. The first plants, weeds, and "second growth," must be adapted to the initial conditions which emphasize openness and exposure. Stability is achieved when the established community can reproduce itself in place, which puts a premium on shade tolerance. Be the succession simple or complex, each new form must compete effectively in the ambiant conditions which in turn tend to change as each stage develops. Each stage is a community which marks how much progress of succession has been achieved. Inasmuch as edaphic effects can be overcome, the concept of succession is also applied. A swamp may gradually fill in with plant growth, a soil may eventually seal a droughty substratum.

Geographers often find persistent alternative communities occupying area which does not display any corresponding environmental distinctions. Examples are the "balds" in the Appalachian Mountains, treeless areas surrounded by trees. Stands of brush particularly seem to defy the advance of forest. Once a cover becomes well established by whatever means, there is the possibility that it may effectively resist invasion by higher forms for protracted periods even though it must eventually yield through attrition. Much so-called succession, especially rapid succession, depends on a mixture with the protagonists developing together. Eliminate a part of the mix and, in some cases, the remainder may hold its own quite well against other somewhat better suited communities. Sometimes just occasional disturbance, as in some Douglas fir forests, is enough to make such a community essentially permanent. On the other hand, a change in the kinds of disturbance may result in a precipitous replacement of the erstwhile entrenched cover. Witness the almost treeless grasslands of Texas where cattle ranchers intruded causing an increase of grazing and a decrease in fire. The whole area changed abruptly into a mesquite woodland just as fast as the trees could grow. But previously, the grassland was a persistent community that apparently had existed for thousands of years. Persistent communities are very much a part of the life pattern of the earth's surface.

It has been repeatedly shown that many plants and animals can live in areas where they have been previously unknown. Particularly does this refer to the many natural and man-induced invasions that have taken place. A natural example is the replacement of many now extinct mammals in South America by holarctic species that has already been mentioned. Man's recent activities have greatly accelerated the process. Isolated regions formerly protected from competition have been brutally altered. Some species have even had great success in becoming established virtually throughout the world where conditions comparable to their homeland are found. American cactus now also grows from Africa to Australia. But African grasses have shouldered aside the native species everywhere in the tropics. If anything, remote Australia has contributed even more migrants. Eucalyptus trees particularly are now established from California to Brazil and Ethiopia. Casuarina marches across many a sandy tropical coast and Australian acacia and paper-bark (*Melaleuca*) have shown themselves to be quite aggressive. European rabbits have made a reciprocal contribution to Australia. Feral horses and masses of tamarisk bring some of Asia into America. There are countless lesser examples including most weeds. In many cases the added elements decisively alter or even replace local communities. The other side of the coin concerns local extinction. Something once present, if eliminated, cannot reappear even when favorable conditions are reestablished. Furthermore what theoretically might prosper does not live where it cannot reach. The limited or unusual communities of isolated islands and lakes are easily understood. Just as isolated may be mountain peaks, basins, or special edaphic sites. Particularly do herbaceous plants, whose seeds are generally efficiently dispersed over great distances, grow to the exclusion of woody plants, less successful spreaders as a rule, in out of the way locations. Consider a meadow on a droughty soil surrounded by towering redwoods or a low tree-line on the southern Andes when the Cascades with a similar environment are crowded with boreal forest species. A local community can only include a selection from that which is available to it.

There is a strong tendency to describe areas in terms of stabilized (potential or climax) communities thus eliminating all troublesome ephemeral aspects. It is useful to think in this way because it allows the identification of fundamental environmental parameters which interact with life. However, what sort of a stabilized community might develop in a specific area is usually unknown and any attempt to make a determination is likely to be very subjective. Even where a more-or-less stabilized community is known to have existed, its reappearance intact is clouded by considerations of soil changes, climate changes, alterations in its parasites, introduction of new species, and even genetic modifications within the community. Dynamic change dominates life communities and the majority that might be encountered from place to place are manifestly ephemeral.

Usefulness of the Concept of Community

The idea of community is specific and it is local so that a study of communities can yield a great deal of information about a local area. It is at this point that ecologists tend to become geographers, that is, the interpretation of ecological knowledge about communities has geographic implications. The more detailed differences in community from place to place that can be adduced, the more insights about local patterns that can be revealed. Application can involve looking forward to future use potential or backward to what has happened in the past.

Potential

More than anything else, the plant life in an area integrates all aspects of the local environment within which it necessarily lives. Being visible it also provides a rapid means of assessing local conditions, what nature offers in each place. A whole series of potential uses of an area, agriculture, stock, forestry, recreation, military, and even urban development can be planned more efficiently with the knowledge that the local patterns of plant communities can reveal. To be sure, indicator species can be and are used in the same way, but the whole community pattern yields more information. With increasing pressure on the land demanding ever more efficient use, a precise appreciation of the quality of the land becomes more necessary. Perhaps that is why much more effort is expended in Europe than in America for ecological mapping.

Planning and management of agriculture is perhaps the most intense exploitation of the land. A knowledge of the local communities was very important to the frontiersman in early America. These pioneers soon learned what the vegetation cover could tell them about the agricultural potential of each area. There is still settlement occurring in the world, particularly in interior Africa and South America where the wise observation of the ecological geography or lack of such observation can spell the difference between success or failure. Each soil factor, including internal drainage, lime and humus content, and many another is reflected by community responses. The same is true of microclimate. Exposure to dessication or pockets of air drainage may be sharply defined by the plant cover. Observation of the plant cover is extensively used in the preparation of soil maps which in turn have wide use in agricultural management.

Less intense production programs than agriculture can profit even more from information about community patterns. For the potential investor an assessment of productivity can be made, and for the producer intelligent management practices can be planned. Manipulation of the plant cover can increase yields of lumber and other wood products as well as improving stock conditions. Forest and range managers therefore begin by mapping the local communities.

The potential of an area is not limited to production. Increasingly the demands for recreation are being felt. Each kind of area offers its own rewards or limitations and much recreation requires careful planning. Biologists need to know about habitat patterns in the study of animal ecology. Military strategists can use many kinds of information. Even urban planning profits from knowledge of land patterns. The list of the uses, great or slight, which may emerge from mapping the vegetation cover, is virtually endless.

Over and above the simple evaluation of environmental potential is the fact that development is not a static thing, nor are man's needs. Nature is in a state of constant flux. Short and long cycles of climate are ranged beside the constant effects of gradation (erosion and deposition), not to mention the tectonic forces. Against nature are man's changing technology and varying objectives, all of which require a continuing reevaluation of the environment, much of it through an appreciation of local community patterns.

History

If communities are dynamic, what is to be seen ought also to tell something about the past. Each kind of disturbance promotes a particular response. Successional communities indicate what sort of disturbance has occurred and likely how long ago as well. The historical geographer can read old patterns from the present cover. For legal reasons it might be desirable to prove when a particular disturbance occurred. An example might arise with respect to a river political boundary where slow migration of the river has a different legal standing from sudden shifts in course. The soil can also record something about the past, and the two together, soil and plant cover, sometimes provide surprising information. The sites of old buildings and old field patterns, gone for centuries, have been discovered by a study of the surface.

The simplest information about the past is mere age of the plants, the initial recovery by individuals. These are the pioneer communities and they often occur as even-aged stands which produce interesting texture patterns to remote sensing. The whole mass called jungle is really only an early stage in recovery of a complex community. Left alone for a few years a jungle becomes a forest.

Succession is usually an orderly process, the first stage of which often is strictly transitional. The pioneer community does not maintain itself but yields to a new community. The steps in this process involve a mixture of old pioneer plants with young successional species. The last pioneers persist as relicts. Sometimes there is a sequence of stages each taking over from the one before, and only when a truly stabilized community has been achieved is there lost the traces of disturbance.

Changes in the soil can be wrought by external forces. One possibility is a change in chemical content such as results from early man's piling of shells or bones or other refuse. The calcium content may be increased or the organic condition may be changed. Local surface configuration may be modified. Particularly is this true where some forms of early farming were practiced with mounds or ridges thrown up across the land. This changes local drainage relationships and therefore alters local living conditions. Whenever living conditions change, the natural communities are likely to be different and hence patterns are produced on the land that can be read by the ecologist. Sometimes, however, a ground observer is too close to see these patterns and many were only revealed with the advent of viewing from the air.

There is a great amount of information that may be derived from the observation of local community patterns. It should be remembered always that the responses to disturbance and to local environment are simultaneous. Erroneous assumptions can be derived from ignoring this interplay. Furthermore, obvious local differences may spring from factors of no concern for a particular objective. That is why the reading of local community patterns requires some care for best results.

Chapter 6

Formations

The aggregate of life forms in an area is known as a formation. Because individuals of the same species can exhibit different forms as a result of a variety of factors, the idea of formation cannot be equated with community, even though a strong relationship might be expected. With formation there is, furthermore, achieved a different approach to the geography of life, one which is largely independent of unique biotic elements because of the fact that any given formation type is composed of different combinations of species in different places. There are therefore greater possibilities for generalization about response to environmental conditions, the kinds of generalizations that have much usefulness in relating different distributions with which a geographer might be concerned. As with communities, formations are dominated by their plant life because in form terms the animals are so much less significant.

The Nature of Formation

Several or perhaps many distinct types of formation can be recognized based upon different combinations of form elements or physiognomy. Certain kinds of formation, such as forest, grassland, or desert, are widely recognized, and their differences are readily identified. Others are less obvious and are not easily defined, requiring an understanding of the elements by which they are constructed. There has been considerable confusion in the literature concerning what should be called a formation. It will be helpful to trace the historical development of the treatment of formation systems after an analysis of the various recognizable formation types has been made.

elements of formation

Each organism and each part of an organism has a form, no two of which are identical. Both popularly and with more rigor, there have been recognized certain kinds of form, the more important of which, at least, should be presented. Sometimes a definition for long accepted form-types is not easily discovered and quite a discussion of just what qualifies as an example of a form-type can be quickly generated. The problem will be considered here in two ways. First, individual form with its many ramifications will be examined. This will be followed by a consideration of certain relationships between forms. The further question of the relationship between taxa and form will also be discussed.

Individual Form

The most obvious way to treat form is to consider each individual organism separately. In effect, the discussion of individual form is restricted to land plants, although to some extent the categories presented here apply also to creatures of the sea. One can talk of the whole plant or one can examine the parts, specifically leaves, branches, and roots. Each category can be subdivided into many types both general and unique.

The Whole Plant

In order to show the range of form which can be distinguished among plants there is presented below an annotated list.

TREE. The most obvious plant form is a tree, but "tree" is tricky to define. A tree is large and rises above the ground on a single pillar called a trunk. Essentially, a human must look up into the branches of a tree and the trunk is composed of wood or some other lignous material. The largest plants are trees.

BUSH. A bush differs from a tree in the proliferation of woody stems rising from the ground. Bushes are generally thought to be smaller than trees but there is much overlap. Many bushes are smaller than people.

LIANA. A plant form that is woody but does not support itself independently being draped over larger plants. A liana can be called a woody vine.

HERB. Those plants which stand by means of turgidity rather than by a woody structure are called herbs. Some are perennial and survive unfavorable periods at or below the ground surface by means of bulbs, tubers, and other storage devices. Others are called annuals and die completely leaving only seeds. The term herb is not rigorously comparable with tree or bush. In the always-wet tropics some herbs become trees, as the banana. Many herbs are essentially bushes and an herbaceous liana is a VINE.

EPIPHYTE. A plant which uses other plants for complete support, not being attached to the soil. There are even tree epiphytes, not to mention herbaceous varieties and epiphytic vines. The term epiphyte refers more to position than to form.

PARASITE. A plant which derives nourishment from some other plant. Those in trees resemble epiphytes, those on the ground (root parasites) are sometimes hard to distinguish from other plant forms.

CUSHION. A colony of tightly packed stems. Moss and lichens often grow as tiny cushions.

CREEPER. A vine-like plant that is prostrate. Some are herbaceous, others woody.

OTHERS. Many more plant-forms can be conceived including such things as *ephemerals* and *succulents* or *bunches* and *rosettes.* There might also be subdivisions of any of the above categories.

What is being emphasized here is the gross structure of the plant. The working ends of the plant are the root hairs and the chlorophyll. The ways in which these are connected is the form and might be developed on the basis of root-leaf connection, such as tree, bush, bunch, cushion, and vine. The internal structure of the connection is also illuminated as in woody, herbaceous, or succulent. The ideas of epiphyte and parasite are disharmonious here and tend to refer to the nature of the root function rather than to form. Raunkiaer devised a system based on the position of the perenniating bud: *phanerophyte* (tree), *chamaephyte* (bush), *cryptophyte* (perennial herb), *therophyte* (annual), and others such as *epiphyte* and *hydrophyte.* These various plant forms are unevenly distributed around the world and combine into characteristic aggregates which have been largely treated descriptively rather than analytically.

Leaves

Much attention tends to be given to leaf form because leaves are the locus of plant food manufacture and also, being displayed to light, they are the most obvious part of a plant. The two characteristics of leaves of particular interest are tenure or persistence and general form.

Leaf tenure primarily refers to the difference between the *evergreen* and *deciduous* habits. These terms are a bit tricky, however, because in a technical sense most leaves are deciduous; that is, they eventually fall away from their branch. Furthermore, there are many plants which drop their leaves and then directly grow a new set, standing bare for but a brief time, and this is sometimes called deciduous. The important idea, then, is that some plants sacrifice their leaves during a dormant period while others have developed protective devices for their leaves enabling the leaves to survive dormancy to function again during a subsequent favorable period for activity. The protective devices involve thickened cuticle, insulating coatings such as wax or hair, the ability to curl, and reduced surface with respect to volume. Leaves having any of these properties are said to be *xerophytic*.

There are just two ways that a plant with perennial shoots can handle its leaves with respect to a dormant period. The leaves can be evergreen and xerophytic or they can be deciduous. There seems to be no real advantage of one of these methods over the other because in seasonal areas where dormancy occurs the two are about equally common. Nor can a convincing case be made for either being better suited to any particular sort of environment. The evergreen plant can function immediately after favorable conditions become established but the xerophytic devices interfere with its best efficiency whereas the deciduous plant must first unfold new leaves thus delaying food manufacture but, once underway, a maximum of efficiency can be enjoyed. The herbaceous habit is obviously a more drastic form of dormancy where the whole aerial plant is sacrificed. The loss of size here is counterbalanced by flexibility.

The general form of leaves involves shape, size, and, as mentioned above, ratio between surface area and volume. Many odd leaf shapes are known, but the vast majority of leaves have very ordinary more or less oval-shaped leaves. Leaf size also tends to concentrate around a norm with larger leaves tending to be divided and smaller leaves aggregated to approximate in effect the intermediate area. Perhaps of more significance is the surface-volume ratio. For larger leaves this ratio is essentially a function of thickness. On the other hand, the greatest volume with respect to area is achieved by a scale-leaf. One can thus speak of three general categories of leaves. The *thin* leaf is most commonly represented by the broadleaf deciduous type and involves the larger sizes. The *thick* leaf tends to be evergreen and is characteristically on the small side. The *constricted* leaf includes needle-like and scale leaves as well as those which are rather small but still somewhat flat. Unusual combinations of leaf character do occur such as deciduous needle-leaves, huge thick slabs, or needle-leaves a foot and a half long. The thicker and more constricted leaves are associated with the more severe environments.

Branches

The function of branches is to unite the roots with the leaves and as a structure to support those leaves. There are many geometrics that can accomplish this purpose and there are differences in the way leaves are displayed. The arrangement of leaves essentially is either on a column or on a plane. With a column there is really no choice in branching structure. The more usual display is on a plane but this in practice becomes more or less of a dome, the branching possibilities for which are many. At one extreme a single trunk may extend unbranched to the very crown producing an *umbrella* effect. High branching may more resemble a *telephone pole*. Dispersed or random branching can be called *dendritic*. Low branching with a dome-shaped crown suggests the *wine glass* shape while basal branching is nothing more than what is commonly called a *bushy* form. The umbrella shape may be the most economical structure with little strength required of anything but the single stem, but it is also the most vulnerable because damage of that stem, even at the very apex, affects the entire crown. The bushy form suffers the least crown loss from the results of stem damage but also involves the least efficient kind of structure. The dendritic form is distinctly the most common and obviously offers a compromise or optimum solution between these two opposed considerations. All these branching forms serve the same function in the end even though they look quite different. Even more distinct variants such as the candelabra shape of the Paraná pine of southern Brazil or the fan shape of the traveller's palm of Madagascar, in spite of their arresting visual impression, should not be particularly singled out as form types.

Roots

Less can be said about roots because they are not visible in most cases. Some roots are deep and there may be a central tap root. Other roots are shallow. The volume or even area occupied by the roots may be less than, equal, or more than that of the crown. Where the roots are more extensive, the individual trees would tend to be free standing rather than crowded together.

By and large, it would appear that only a few elements of individual form are useful in sorting plants into groups. Leaf size and shape together with branching structure crowd around a middle position that dominates over the other possible character categories. The best results of form differentiation seem to come from the gross structure of the plant, the surface-volume relationship of leaves, and the general life or tenure of plant parts.

Relative Form

Rather than take form on an individual basis, it is more productive to consider form-elements in relation to one another because form elements hardly exist in isolation. Where individuals of many forms congregate the most simple and obvious relationship is size. More important, however, is the position with respect to others achieved by each form type. Other relationships might include density or even tenure which was discussed above.

Size

Within as well as among the various kinds of form there are individuals of different size. Trees reach the greatest sizes but there are big trees and little trees and others in between. If mixed, the bigger trees must overshadow the smaller trees which in turn would rise above most bushes. Size is an aspect of the differentiation of gross forms even though there is overlap. Size has been used as a primary vegetation formation factor but always in an arbitrary way. Very little can be deduced from size alone except that larger plants tend to shade smaller ones. If the smaller ones can outlive the larger ones or are more tolerant, they may yet replace them in spite of lesser stature.

Position

Instead of size alone, the relative position of the leaves, the locus of activity, can be considered. Because the biosphere is a surface, relative position essentially involves layers or *synusia.* There are three basic layers to the plant cover:

CANOPY. Where the tree crowns exposed to the sun are aggregated the result is called a canopy. Within the canopy, other forms than trees may take their place, including vines and epiphytes. It must not be supposed that the canopy is necessarily an even blanket of foliage, however, because trees of different sizes may mingle in greater or lesser amounts. Individual trees whose crown rises above the neighbors are called *emergents.* Where taller trees are numerous there may develop a two-layered canopy. The elevation of the canopy is variable; some canopies soar over 200 feet above the ground, others crouch at no more than 40 feet.

UNDERSTORY. Trees completely submerged under the canopy comprise the understory. Included are juveniles of the canopy trees themselves, poor competitors which rarely achieve the canopy, and understory obligates unprepared to survive open exposure. The understory may press intimately against the underside of the canopy even though clearly being differentiated from it as a layer of a different aspect. Further, trees not in the uppermost layer of the understory are sometimes considered to form a second understory.

GROUND COVER. Although the understory could theoretically reach to the ground, the ground cover is a different phenomenon. It represents the renewable growth that can replace itself in one season of growth. Involved are the herbaceous plants together with the sprouting shrubs and creepers which grow at the same level. The size of the ground cover is determined by the height to which the most vigorous available participants can reach and all others which can survive together with them. Some ground covers rise only a few inches, others stretch to several feet, that is, even above a man's head. There are really two kinds of ground cover, the one which grows in the shade and the other which grows in the sun. The sun-growing ground cover may include its own understory plants, necessarily also renewable. Some ground-cover plants may take several years to accumulate enough root support to attain full size, but thereafter they can produce sprouts as large as their competitors. Plants which by increments of growth rise above the ground cover usually continue well above that level leaving the ground cover as a distinctive part of the overall cover.

Obviously in these three layers of plant cover are accommodated all the different plant forms discussed earlier as well as all differences in size. As will be seen below. the elements of cover become basic to the development of complete formations. The possibility also still exists with reference to the characteristics described above, to include leaf conditions in the final aggregate which is formation.

Taxa and Form

It has already been stated that taxa are variable as to form, but this subject merits elaboration. There is always a tendency to equate taxa and form because each taxon has characteristic form manifestations with only a limited range of variation. As true as this may be, the fact remains that form is not defined by taxa.

The first difference between form and taxa is the fact that many taxa have essentially the same form. The announcement of a particular kind of form therefore does not reveal which of the many kinds of taxa may be involved. The converse, the possibility that a characteristic taxon may be used as a form-type, has however been applied. Thus, grass is dominant among herbs and a grassland is understood to include other herbs, although it might be more accurate to speak of an "herbage."

The other and decisive difference between form and taxa is that any taxon broad enough to cover any appreciable area will generally manifest a significant range in form. Precisely those species which are widespread are variable, a factor basic to their becoming widespread. But at the world scale, species are insignificant and, as soon as one has recourse to larger taxonomic elements, much variation is inevitably encountered. What, after all, is the form of an oak? There are small bushes and large trees, not to mention evergreen and deciduous forms of oak. The radiation of a larger taxon into a variety of habitats is in part accomplished through plasticity in form.

formation types

By aggregating the various form elements which have been discussed, a formation is defined. If specific formations can be recognized, it is implied that distinctions can be made. In spatial terms this means that vegetation cover is not a continuum and discontinuities exist. Thus, attention will be given in defining each formation type to its boundary relationships with neighboring formations. Three categories of formation can be distinguished, the first of which will be called basic. The others are disturbed and special formations.

Basic Formations

Widely distributed formations not significantly altered by disturbance are here grouped as basic types. These formations are differentiated by the proportions of each cover type which are present. That is, canopy, understory, or ground cover can be either continuous, discontinuous, or absent. By continuous is meant that the elements are in contact with each other forming a matrix. Discontinuous, by contrast, means that elements are isolated from others of like character. There are four basic formation types or five if one includes barren areas as a type.

Rainforest

The most luxuriant aggregation of plants is a rainforest. Not only does a rainforest have a continuous canopy but it also has a conintuous understory, which no other formation has. There is no ground cover. The profusion of species so often present has led to the recognition of subdivisions of the canopy and understory, from a double canopy to even a triple canopy (that is, understory layers are being called canopies). It is the continuous understory which both defines a rainforest and accounts for the profusion of tree species.

The rainforest canopy is complex and xerophytic. Often the commonest tree species of the canopy does not surpass five percent of the total but, considering canopy trees alone, a single dominant is not unknown. Usually emergents, huge trees sometimes over 200 feet high are scattered above the canopy which may be 100 to 125 feet above the ground. Grasping for a share of the sunshine are various epiphytes and lianas. Favored as it is by the direct rays of the sun, the canopy is also exposed and leaves in it are necessarily tough and xerophytic.

By contrast, the understory is a highly protected place. Although allowed only subdued light, the understory enjoys high humidity and steady temperatures together with little air motion. The bigger and smaller trees which live there usually total hundreds of species and include plants such as palms and tree ferns with huge compound leaves. The prevalence of "drip tips" on leaves of all kinds of plants suggests the moistness and the danger that wet leaves have for a plant, a place for fungus and other pests to flourish.

Wherever the understory is protected from decimation by excessive cold or drought, there can a rainforest prosper, but several varieties can be singled out. On the drier margins there is a structure variously called a "seasonal rainforest" or "dry rainforest," among other names, which has a reduced complexity. Here the canopy trees may be more or less deciduous during a dry season and the understory may show signs of stress. In cool areas such as mountain zones or mid-latitude maritime areas there may develop a "temperate rainforest" with not only a reduced complement of tree species but also with leaves reduced or even constricted. Such a stand might better be called a "sclerophyll rainforest." Neither of these varieties is sharply marked off from adjacent typical rainforest and both continue to display the uninterrupted understory.

Margins of the rainforest can be remarkably sharp. Seen from the outside it may look like a wall of leaves. Obviously, the environment is not sharply divided and rainforest borders may be less distinct, though seldom indistinguishable. What is happening is that, with fluctuations in the environment or in disturbance, the complex which is rainforest is able to advance or is forced to retreat. Submerged non-rainforest trees mark an advancing front and dead rainforest stems tell of a retreat. Either the riot of understory species can endure or they are eliminated, in which case another formation is found.

Seasonal Forest

A continuous canopy defines a forest and where there is no more than a discontinuous understory this becomes a seasonal forest. There is normally a continuous ground cover in a seasonal forest but this is not always present. Many names have been given to the combination being called seasonal forest here, some representing varieties and others such as "lighter tropical forest," "temperate forest," or "boreal forest" involving extraneous environmental references. The word seasonal may be awkward with respect to the defining character of no continuous understory, but it refers to the dormant season which occurs and which accounts for the relative lack of tender understory plants.

The canopy of a seasonal forest is simple, being dominated in most cases by only one, two, or three kinds of trees, although a "kind" here may represent more than one related species such as two species of fir. As a result, individual stands of seasonal forest can usually be named in terms of dominant trees as an oak-hickory forest. Under the most favorable conditions a combination of four or more important trees may develop as in the *mixed masophytic forests* of China and the southern Appalachians. Epiphytes and vines may occur but, if so, tend to be restricted to a few varieties.

The understory of a seasonal forest has a preponderance of canopy tree juveniles. Trees of the understory, such as dogwood, are generally quite disperse. At times there may be almost no understory but, normally, juveniles would be present. Ground cover plants obviously must be shade tolerant and include a random selection of herbs and shrubs often very specific for some microhabitat in the forest.

The seasonal forest can be divided into lighter and denser phases both of which extend the entire width of the seasonal forest zone. Lighter seasonal forest has a rather open canopy which allows a great deal of indirect light into the forest, whereas the denser type of forest produces a deep shade in its interior. These two types are not sharply differentiated and the lighter one may be only an incompletely developed stand. On the other hand, the lighter forest tends to be associated with strong dry seasons, such as in the tropical seasonal forests. But these areas are chronically disturbed if the forest survives there at all. The relationship between lighter and denser seasonal forests is as yet not fully understood.

Normally the dominant species of a specific seasonal forest all grow to nearly the same height. What this height is varies tremendously from one stand to another. It sometimes happens, however, that trees of distinctly different stature become associated, that is, a taller tree species does not achieve a complete coverage. The result is a "layered forest" with a divided canopy. The taller trees are, in effect, emergents. Among the relatively few examples are some forests of northern Australia.

Leaves of seasonal forest trees are either deciduous or they are evergreen and xerophytic; normally there is a mixture. Such a mixture can be called a *semi-deciduous forest* or a *mixed forest* (although *mixed* could have other meanings). Because only one or a few dominants characterize a seasonal forest, it is fairly common that these dominants not be mixed; that is, the stand may be all deciduous or all evergreen. These are special cases of the seasonal forest and their distribution is randomly related to the more common mixed variety. The contrast between evergreen and deciduous is so obvious that the distinction is usually mapped even though it probably is not of major importance. As environments become more extreme, leaves become more constricted even to needle and scale shapes, particularly for evergreen examples. This leads to the designation of *needle-leaved forests,* largely a distinction of degree. By combining all of the possible differences, a large number of subdivisions of seasonal forest can be and often are made.

The margins of seasonal forests, even in the absence of general disturbance, are rather sharp. On one side is the rainforest whose boundary with the seasonal forest has already been discussed. The opposite boundary of the seasonal forest is essentially similar. Evidences of advance or retreat mark either side of a boundary that may be no wider than the reach of the branches of the last tree and is rarely more than a fraction of a mile wide. Where the canopy growth is stabilized there is seldom less than an 80 percent cover, while beyond the seasonal forest there is seldom more than a 40 percent canopy cover. Evidently, where more than 40 percent cover is achieved there is enough protection in the small openings for forest juveniles to grow and to close up the canopy. The seasonal forest is an unquestionably distinct formation.

Woodland

A continuous plant cover with a discontinuous canopy defines a woodland. Nothing much of an understory could be expected under such conditions, and the ground cover would be mostly of the exposed type. Terms such as *scrub forest* and *thorn forest* are often applied to woodland but represent an undesirable over-use of the term forest.

The canopy of a woodland has most of the characteristics of that of the seasonal forest. Dominants are few and may be either evergreen or deciduous. Constricted leaves are quite common as are thorns. Canopy trees are necessarily isolated and thus their branches surround the stem and form a skirt down at least to the ground cover. Actually, in most woodlands the trees tend to clump two and three back to back, although never would a tree be completely surrounded by neighbors. Trees adapted to growth in full sunlight cannot ordinarily survive complete immersion in shade. As was noted above, a maximum coverage of 40 percent seems to apply to the woodland canopy. A minimum coverage of about 25 percent also obtains and this seems to result from the spread of roots. Seldom does a tree's roots average more than twice the reach of its branches so if the crown coverage falls below one quarter of the total area, openings appear in the root zone where new tree individuals would have an opportunity to fit in. Woodland trees can be quite large but this confers no particular advantage and it is not surprising that woodlands are usually rather low.

Ground cover in a woodland is a mixture of herbs and competing bushes whose relationship is much like that between evergreen and deciduous trees. Sporadic patches of bare ground may occur. Underneath the woodland trees there may be a different, shade-loving ground cover. It is of interest that woodland tree juveniles seldom grow under their elders but rather find a start either at their edge or under a ground-cover clump.

Aside from the recognition of leaf differences, the woodland can be divided into two phases based on stature. Plants that rise above the ground cover may yet be so low that they only merit the name bush. Such is the case with sage brush, for example, and the result could be called a *brush woodland* as opposed to a *tree woodland.* Where smaller and larger plants overtop the ground cover a two-layered condition exists, but because the canopy is discontinuous the result is not particularly noticeable. Brush woodlands appear to be related to the most severe woodland environments, that is, both cold and dry together.

Limits of woodlands are difficult to distinguish, and there are two reasons for this. In the first place, woodlands are the most prone to disturbance of the basic formations with often only fragments surviving. The other reason derives from the fact that their matrix is the ground cover which is by definition variable. In bad times the ground cover may wither leaving the whole plant cover discontinuous. Yet with the return of better times the ground cover is eminently suited to closing in again. Nevertheless, the woodland edge is really rather sharp because, in the good years, the invaders from the desert which may have gained a foothold are crowded out. There can, therefore, be found a woodland frontier much like those of rainforest and forest which may advance or retreat slowly.

Desert

Where only a discontinuous cover of the total plant content is achieved, the formation is called a desert. As in a woodland, clumping of plants is a common feature and juveniles rarely germinate on open bare ground. The total cover of the desert formation rarely exceeds one-third, even where it lies close to the woodland boundary, but after a heavy rain, desert ephemerals may spring up and produce, for a few weeks, a very un-desert-like cover.

There is no reason why desert plants might not include trees and indeed trees do grow in some desert areas. Giant cactus and palo verde trees in Arizona combine with other bizarre forms to produce what may be the most striking kind of formation. Aerial size confers no particular advantage in a desert and, where conditions are more severe, only small plants huddle near the ground. This suggests two varieties of desert, a *tree desert* which grows near the desert margin in warmer zones and a *brush desert* elsewhere.

In a few areas of the world the ground is completely bare of plants and such an area could be called *barren desert* or it could be counted as a fifth basic formation. Aside from bare rock exposures and naked ice, permanently bare surfaces only occur in almost waterless areas such as northern Chile, parts of the Sahara, and northernmost Canada. Whether any sort of boundary with plant deserts might exist has not been determined.

Disturbed Formations

The alteration of the vegetation form-complex in an area, when of a significant amount, produces new formation types which can be called disturbed formations. It is not always possible to determine whether a particular combination of forms developed first and the disturbances came later rather than that the formation was produced by disturbance. Furthermore, what results from disturbance in one place may be paralleled by lack of availability of certain forms elsewhere. Therefore, the identification of disturbed formations can be subjective and is definitely debatable. Certainly the woody plants which, by increments of growth, attain greater size, have an investment to lose from disturbance. Ecologists argue long about the role of disturbance in form but no attempt to develop the full argument will be made here. Briefly, by disturbed is meant those formations whose woody content is less than in the basic formations. There are three such formations.

Grassland

A continuous cover of herbaceous plants (usually dominated by grass) is called a grassland. Little or no woody plant growth is present. Almost all known grasslands have a history of fire which tends to prevent the development of larger woody plants if these are available to the area. Frequently, grasslands have sharp boundaries with woodlands, forests, and even deserts, formations which differ absolutely from grasslands by the presence of larger woody plants.

Several variations of the grassland formation are often distinguished based on secondary structural characteristics. Tall (prairie) and short (steppe) grasses are mapped, and certainly there is a gradient from taller and denser sod to shorter and sparser cover but, as in the Great Plains, where the division is to be made is difficult to specify. Some grasses produce a tangle of rhizomes which is called a sod and others grow in bunches, leading to the distinction of *bunch grass* for some areas. Bunches crowd together to make a continuous cover so the net result is the same.

Savanna

Scattered woody plants in a grassland constitutes a savanna as the term is usually used today. Isolated woody plants are relatively safe from endemic fire because grassland fires sweep along the ground and are not very hot, rarely burning out the crown of a woody plant. This is true because where bushy plants grow down to the ground either the herbaceous growth is too thin next to the bushy leaf growth or the leaves are too lush to ignite. Free standing trees may experience some scorching of their lower branches but their trunks are usually too tough to suffer any serious damage from a surface fire. It takes a nearly continuous canopy to sustain a crown fire.

Savannas cannot always be distinguished from other formations. All degrees of woody growth from those of the basic formations to none at all can be found and, if disturbance is the reason for a savanna formation, this is what should be expected. Normally a savanna would have trees or bushes so widely spaced that their root areas would not be in contact. That is, the woody plants in a savanna do not have a continuous root zone and this suggests a crown cover of less than a quarter, often much less. On the other hand, no one has ever established how much woody growth distinguishes a savanna from a grassland.

Because of the range of density and size of woody plants that may be present, various savanna types have been distinguished. There is an obvious difference between a *tree savanna* and a *bush savanna*, although no sharp dividing line can be made. Even grassland could be considered a variety of savanna. Where the ground cover in a woodland is completely herbaceous it is called a *woodland savanna*, more a kind of woodland, perhaps, but also a savanna.

Brush

A continuous ground cover dominated by woody plants can be called brush, although the term is also applied to what was called above a brush woodland. Continuous brush has earned many distinctive names such as chaparral, maqui, and matorral. Ground cover plants include sprouting bushes which are quite woody so that grassland and brush are really only two varieties of a continuous ground cover, a cover where incremental plants are absent. The relationship between brush and grassland ought to be like that between evergreen and deciduous trees in a seasonal forest but, actually, brush is much the less common. In fact, brush is so sporadic and limited in distribution that it cannot effectively be included on maps of very large areas and it is relatively common to combine it with woodland. Brush cannot be any more clearly distinguished from woodland than can savanna and it also can grade into bush savanna, but brush differs sharply from forest, grassland, and desert.

Special Formations

Those formations which are, by their very nature, of a local character can be called special. Some may not differ in their general structure from more widespread types, particularly those which involve some kind of standing water. One could probably distinguish a vast number of special formations by recognizing more detail, but only those more widely referred to will be listed here.

Elfin Forest

On exposed ridges, mostly in the humid tropics, there grows a short dense forest which essentially is a canpoy with no space underneath. Such places are characteristically bathed in clouds and dripping wet but are subject to sudden interludes of bright dessicating sunshine and sometimes wind. The result is a stunted growth of thick-leaved trees, some of which may grow to great size in protected localities. An amazing accumulation of epiphytic growth is present, from moss and lichens to the most highly evolved groups. Although immensely beautiful, penetration of an elfin forest is a strenuous athletic event.

Bog

Cool moist areas where peat accumulates on the surface of the ground is called a bog. Muskeg is a synonym. Usually there are stunted trees present. It might be argued that bog is more of a soil condition than a plant formation because it is only the accumulation of incompletely decomposed organic material over the ground surface which defines the bog.

Swamp

A forest standing in water is a swamp. Surprisingly, there may be a well-developed ground cover, partly because the water level is usually subject to fluctuation and there may even be a season without a water cover. A swamp is not really a distinct vegetation formation and the plants that grow there often extend into better drained areas.

Marsh

A wet grassland is called a marsh. There are salt marshes and freshwater marshes, the plants of which are distinct. Like a swamp, a marsh is not really a distinct formation.

Mangrove

Just a few tree genera can survive in saltwater tidal conditions and these only in the tropics. Coastal saltwater swamps are called mangrove. Special organs such as knobs, knees, and stilts are particularly developed among mangrove species, although also found in other swamps. Although easy to spot in terms of its habitat, mangrove doubtfully merits distinction as a formation.

As a matter of fact, unless one is prepared to include soil and water conditions as elements of form, all but elfin forest among the above widely used special formation types might have to be rejected. This is not to say that soil and water conditions are not important for life. Bog and marsh and mangrove are significant habitat types but are doubtful formation types.

historical development of formation systems

Current in many atlases and text books are systems of vegetation formation which differ substantially from the one presented here. It is of some interest to explore briefly the development of thinking concerning vegetation in order to appreciate the differences which may be encountered.

Early Systems

During the nineteenth century enough was known about the environment that general schemes embracing the whole world became possible even if detailed maps were not. The most famous attempt was by De Candolle who proposed a two-way vegetation breakdown. There were four temperature categories, roughly tropical, subtropical, subpolar, and polar, beside which were three additional arid categories, desert, steppe, and woodland. Also influential was the system of Schimper which appeared at the end of the nineteenth century. Schimper adhered to the classical three-way division of temperatures inherited from the Greeks but for his warmer two zones he proposed no less than five moisture divisions: rainforest, forest, woodland, grassland, and desert.

Maps produced before the twentieth century were rather tentative and imprecise. There was a tendency to try to organize the categories according to a system but usually a few disharmonious types were included. A major problem stems from the fact that the two major recognized systems, although similar in having a temperature sequence and a moisture sequence, differ in detail. De Candolle has four temperature divisions, Schimper three. De Candolle has a total of four moisture divisions and Schimper five. Many partial compromises were attempted such as having just one rainforest and two steppes and deserts but three forests. Woodland seems to have been the most perplexing category to place.

In the meantime, more detailed maps were being prepared of parts of the world, usually continents. Knowledge of the patterns within some areas was available while other parts of the world were still to be explored. Such regional maps could include unique categories such as prairie, taiga, and caatinga that were not rationalized with respect to other parts of the world. It sometimes happens, even today, that world maps are prepared by simply combining a set of regional maps, with the result that a lot of unsystematic parochial categories are included. On the other hand, local categories, such as taiga, have been elevated to a general status by combining with other similar types.

Later Development

Several trends emerged as the input of information on vegetation expanded. The first systems were, after all, combinations of climate and vegetation concepts and so it is not surprising that climatic thinking had an impact on the development of vegetation maps. So too did ideas stemming from local considerations. A local category of vegetation might, if handled by a botanist, be identified by its taxonomic dominants or by particular form characteristics as by some generally applicable trait shared with different plants around the world. These considerations might then be expanded as distinctions in a world system.

Climate has had a powerful impact on the understanding of vegetation. From the start, climatic zones were taken as vegetation zones. Not only were grasslands divided between tropical and middle-latitude varieties, but this was validated by the assertion that tropical grasslands were savannas and middle-latitude grasslands were clean. Equally incorrect is the suggestion that tropical woodlands are deciduous and that middle-latitude woodlands are evergreen. More important, however, was the virtual restriction of middle-latitude woodlands to the dry summer subtropics, a manifestly distinct climate, even though it has no more of a monopoly on woodlands than does any other semiarid part of the middle latitudes. The tundra was also certified as a vegetation type although it represents a variety of formations, none unique, but all without trees. Tundra may make sense as a climatic type but does not deserve to be labelled as a unique vegetation type.

The equation of sub-polar forest with the taiga or northern coniferous forest has led to various results. One is the taxonomic separation of conifers as a major category of vegetation which, considering the variety of conifer forms, is indefensable. Other taxonomic groups, such as palms or bamboo, have also been singled out. Rather than conifers, the needle-leaf form can be recognized. This had led to a whole series of leaf categories such as the contrasting forms with needle-leaf and the difference between evergreen and deciduous. Needless to say these do not correlate with taxonomic categories nor do they correlate with many other possible factors. It is true, however, that leaf characteristics are valid aspects of form. By assembling all of the proposed factors, a vastly complex map can be produced that is sadly weak in system.

Contemporary Schemes

Vegetation renditions in current use involve several contrasting results. Some are essentially a set of types whose origin lay in the old idea that there were natural regions and that, therefore, one could treat climate and vegetation simultaneously. Unfortunately this origin is forgotten today but it is standard practice to compare the resulting climatic and vegetation patterns to show how closely related they are. This is supposed to demonstrate cause and effect, but it really demonstrates that the two patterns were carefully engineered to be the same. A few attempts to produce independent vegetation maps have been made and the results are becoming available.

The more or less conventional vegetation map has about a dozen categories. These include rainforest, savanna, desert, grassland, Mediterranean scrub, middle-latitude forest, coniferous forest, tundra, and ice. Other categories not always used are tropical semideciduous forest, thorn forest, and mixed forest. Still other categories sometimes appear. In terms of vegetation this set of categories is unsystematic and illogical. It does somewhat better if considered as a set of major habitat types.

The first really systematic vegetation map was produced by A. W. Küchler and has appeared in a number of places. There are rather numerous categories distinguished by leaf type, plant size, and overall cover. Grass is also a category. Such a system can be quite informative or even can contain too much information. Its main limitation is that some of the distinctions, such as minimum height of three feet, are rather arbitrary and one must accept the idea that leaf type is of major importance.

The vegetation categories outlined earlier in this chapter and also elsewhere are another attempt to be systematic. By identifying cover elements, canopy, understory, and ground cover, the vegetation is classified by distinct elements that are not arbitrarily taken. Research has shown that real discontinuities exist between continuous and discontinuous aspects of each. A relatively few major categories can be presented, for each of which there may be some varieties.

The Genesis of Formation

Being an integrated complex, the formation ought to relate in rather specific ways to the whole of its environment and to the elements of which it is composed. The geography of formations will display these relationships clearly and thus suggest how the different vegetation formations are generated. With a map in hand (Map 6-1), the more pertinent factors can be considered, particularly climate, disturbance, and availability patterns. A more intimate look is necessary to see the relationships with soil and biotic patterns.

climate

It is generally understood that climate has a major effect on plant form, a situation which is apparent also in the above discussion of form elements. Xerophytic and deciduous leaves are devices whose utility relates to seasons climatically unfavorable to growth. The ground cover owes its existence, in effect, to climatic seasons. Climate is a continuous set of variables which can be rationalized into zones and, indeed, the early vegetation systems were based on climatic zones. It is no surprise that any world map of vegetation will show such zones.

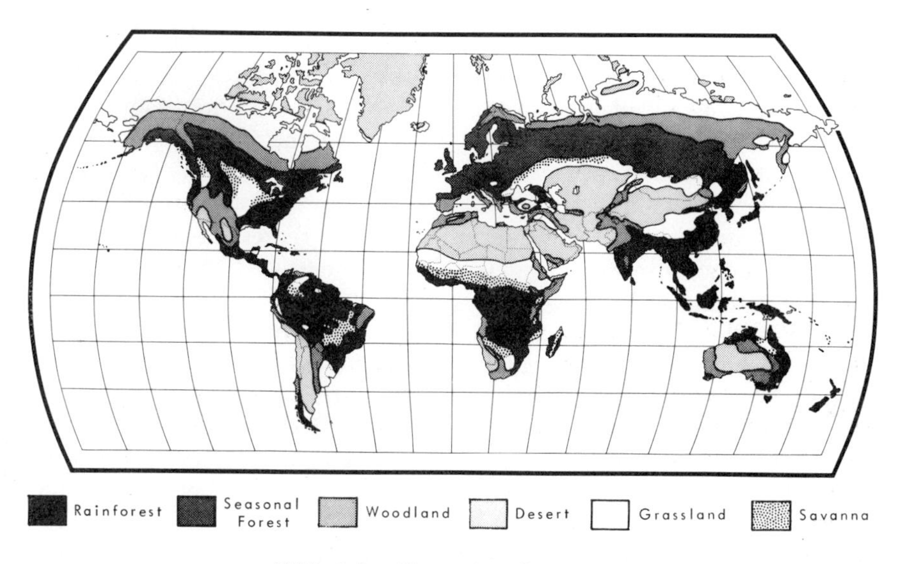

MAP 6.1. Vegetation formations.

The basic vegetation formations presented here form a zonal sequence from rainforest to desert. The rainforests have their greatest development near the equator, appearing also along many tropical coasts and mountain slopes as well as in a few middle latitude areas. These are precisely the areas where no severe dry or cold seasons occur, seasons which would overcome the riot of tender understory plants that make the rainforest distinct. Distributed irregularly along the margins of the rainforest areas is the seasonal variety which shows some reaction to seasonal stress. The sclerophyll rainforest flourishes in cool, moist areas, in tropical mountains, and in middle-latitude marine regions that lack a hot season as well as an unfavorably cool or dry season. Temperature and moisture are both involved in limits of rainforest growth.

Seasonal forest is scattered through the tropics in areas adjacent to the rainforest particularly in Central America and Colombia, India and Southeast Asia, southern Brazil, Africa south of the Congo, and northern Australia. A broad belt in the middle latitudes also harbors this formation. Seasonal forests stretch from Europe to the Amur region and from the Pacific to the Atlantic coast of Canada as well as in eastern United States, eastern China, southern Chile, and southeastern Australia. All of these areas enjoy a distinct moist season without frost, adequate to account for forest growth, but all experience an unfavorable season requiring dormancy and eliminating most understory life. All of the tropical seasonal forest and some areas elsewhere have a dry season while the more severe season in most of the middle latitudes is the winter. Lighter seasonal forest tends to occur where dry seasons prevail but may also expand with the help of disturbance into more moist areas where denser seasonal forest grows. Seasonal forest develops in areas with either cold or dry seasons.

Woodland is found as an interrupted fringe beside seasonal forest zones, particularly in broken terrain. It is well represented in the subtropics as in southwestern United States and northern Mexico, through the Mediterranean and beyond to western India, east of the Andes in subtropical South America, and in west and central Australia. More isolated occurrences lie in central Chile, the southern tip of Africa, eastern Brazil, and eastern Africa. Finally, there is a long strip from Alaska to Labrador and northern Sweden to Kamchatka. What these areas have in common is a distinct but limited growing season. For some, such as the Mediterranean region, this is in the spring between winter and a dry summer. For most of the remainder, this is in the summer but the balance of the year varies from drought in the tropics to bitter cold in the arctic.

Desert occurs as blocks sometimes surrounded by woodland. Most desert areas have well-known names such as the Sonoran Desert, the Sahara and Arabian Deserts, Turkestan and the Gobi Desert, the Atacama, the Namib, and the Great Australian Desert. Patagonia also includes some desert. High latitude deserts, particularly in northern Canada, are less well-known except for those covered with ice such as Greenland and Antarctica. Desert areas all share the lack of a distinct or reliable growing season all except the ice deserts suffer from extreme drought. There are some plantless deserts in a few places. Tree deserts enjoy the least severe desert conditions.

Grassland also has a zonal distribution which tends to parallel that of the woodland. Grasslands surround most desert blocks, particularly on rather level terrain, separating desert from seasonal forest. Woodland intervenes, however, in many places on either side of the grassland zone. Shorter grasses lie next to woodland and desert, taller grasses grow beside forest zones. The great grasslands include the North American Great Plains, the Asiatic Steppe, Mongolia, Tibet, the Sudan, the Venezuelan Llanos, the Argentine Pampa, the South African Veldt, and the grassy Tundra. Smaller areas also occur. The climate of grassland areas overlaps that described for woodland and for seasonal forest but the woodland type predominates.

Savanna grades into grassland but has, to some extent, a zonal distribution, lying either beside rainforest or beside seasonal forest. The North American and Russian savannas or parklands are scattered between seasonal forest and grasslands. In Africa, a belt of savanna separates the rainforest from the Sudan grassland. In Brazil, savannas lie between rainforest and seasonal forest. Elsewhere, savannas are more scattered but have similar positions. Essentially, the savannas all intrude into the same climatic zone as seasonal forest; at least this is true for tree savannas. For the most part, bush savannas are scattered within the grassland zone.

Brush is of too sporadic occurrence to display much of a zonal character. Briefly, brush is associated with woodlands and on world maps these two are usually combined.

The basic vegetation formations are primarily zonal in distribution in response to certain climatic parameters, the dominant factor of which is length of the favorable season for growth. Grasslands and savannas interrupt this zonal sequence while at the same time partaking of some of its qualities. If tree savanna be combined with seasonal forest and grassland combined with woodland, the result is a simple four-zone pattern that forcefully presents the facts of plant growth potential (Map 6-2).

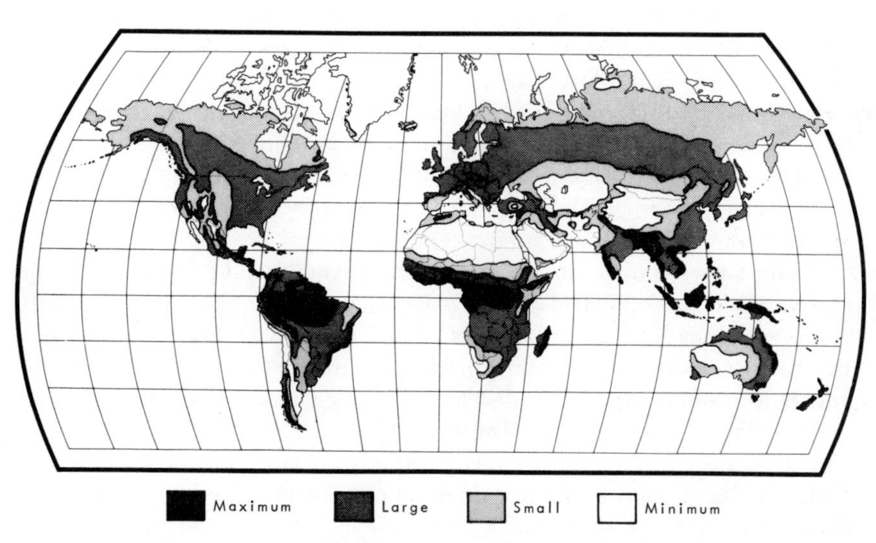

MAP 6.2. Plant growth potential zones (compare 6.1).

Disturbance

Where the very form of life in an area has been distinctly altered by disturbance, one can speak of disturbed formations. There are various kinds of disturbance that may cause these effects and there are several kinds of resulting formations. Obviously, the response to disturbance is likely to be rather dynamic and the resulting patterns complex.

The most widespread and damaging kind of disturbance, other than actual removal by man, is fire in its various forms. There are crown fires in the canopy and surface fires in the ground cover as well as slow ground fires in the soil. Floods and windstorms can be devastating, while disturbance is also caused by animals from locusts to elephants. In all cases, the greatest effect obtains for the larger woody plants which compete successfully by achieving superior size through increments of growth and thus have an investment to lose. Fire and other disturbance is not normally directly beneficial to herbs and sprouting shrubs but indirectly these forms obtain a competitive advantage over the other forms.

Certain very specific things can be said about the distribution of disturbance and its effects. Animals probably have their greatest damage in more severe climates where recovery is slow or, during dormant periods, not at all. The prevalence of thorns in such areas shows the advantage to be gained where excessive browsing is discouraged. Storm damage is probably greatest in forest areas, but above all is the importance of fire. Desert areas do not have sufficient growth to sustain the spread of fire. Rainforests are too wet. Seasonal forests are vulnerable and woodlands are wide open to fire. Terrain also seems to be a factor. Fires sweep across level areas but tend to remain local in broken terrain. Even where fire burns in broken terrain, it causes less damage because of the irregular and often sparse growth where soils may be thin and outcrops are scattered.

Complete removal of larger woody plants would leave only the ground cover and particularly herbaceous plants. A place where woody plants persevere in abundance is called brush, as has already been stated, and where herbaceous plants dominate, the cover is called a grassland. Referring to the distribution described above of grasslands and brush, it can be seen that these patterns are entirely consonant with the possibility of disturbance. Often, grassland and brush alternate with woodland even to forming a mosaic, with grassland predominating on level terrain.

The presence of scattered trees or bushes in a grassland results in savanna. Isolated hardy trees can survive quite well the results of a surface grass fire which rarely has the power to rise into the tree crown. Such trees might either be the survivors of a former forest or invaders which have managed to become established. That tree savannas might be most related to forest zones is understandable but there is no reason to suppose that this would always be so. Woodlands have trees and clean grasslands might, and in fact do, occur among forests. Savannas also form mosaics, the other members of which are forest and grassland.

Not only does disturbance sharpen the boundaries between formations but can promote the extension of one into the territory of another. The margin of a more sensitive formation tends to be displaced by some action, such as fire or windstorm, which may be very localized. Alternately, openings may be carved by such pressures within a stand allowing a more hardy formation to become established.

Disturbance is the rule in most parts of the world, even where farmlands are disregarded. It is easy to overlook areas in obvious stages of recovery, old field regrowth, "second growth," and the like, but the fact is that some covers, once established, can be quite persistent. Before the vegetation cover of an area can be fully explained, account must be taken of these persistent or tenacious stands. In effect, what is being said here is that grassland, savanna, and brush tend to be persistent disturbed formations whose distribution is a function of past occurrences. Some experts would argue that a substantial proportion of these formations did not originate from disturbance and that fire, which occurs, is quite secondary. It is difficult to imagine how climate alone could account for dominance of a ground cover over larger forms but the following factors undoubtedly can explain stable examples.

Availability

It is manifestly obvious that what is found in an area is a selection from that which is available to that area. What theoretically might be able to compete is irrelevant if it is not available. There are two general ways that the plant content of an area might resemble disturbed formations as a result of availability factors. Any kind of isolation, an island, a mountain peak, a unique pocket of soil, might not have been reached by all the forms that potentially could flourish there. Alternately, forms may have been eliminated by disturbance, putting an area out of reach for former and potential inhabitants. It is no use arguing that Kansas grasslands ought to support plants such as pinyon and ironwood or scrub oak when these plants are nowhere in the vicinity to supply propagules. Grasslands and brush survive in many places as stable formations in the absence of larger plants to replace them. Man has in recent times greatly increased the availability of many exotic plants and animals.

Soil

Those soil conditions, not related generally to climate, tend to be rather local in pattern and in effect on plants. Soils can extend the range of a vegetation formation beyond its general body by counteracting the general climatic environment for the better or for the worse. Edaphic conditions can thus produce irregular and sharp boundaries which resemble the effects of disturbance. Soil may also exist in pockets not suitable for the local plant cover and beyond the reach of the usual flora of such conditions, an availability problem.

Biota

It seems reasonable that formations might coincide with and be the result of biotic distributions. That such is generally not the case results from several factors including variable species distributions and aspects of community patterns.

Individual plants species and even dominants do not necessarily change where the formation changes. Some species may be sensitive to the tensions near the formation boundary and keep well back. Many eastern forest trees do not grow to the western limits of the forest in North America, for example. Other species may be hardy and extend beyond the formation of which they are primarily a part. Woodland mesquite also grows as a less prominent plant in the desert. Still other plants may be boundary species found only in the vicinity of a formation boundary, such as Joshua trees. Other plants may find special edaphic conditions to extend their range. Yellow pine woodland sometimes grows on difficult soils beyond the yellow pine forest. Species distributions simply cannot be equated with formation patterns to find formation boundaries.

Communities, by and large, are limited to one specific formation type. Accurate community maps help to find formation limits but must be used with caution. Probably the biggest problem here is the effects of disturbance which fragment the boundaries. However, there may also be boundary communities and other community effects related to the distributions of individual species. Actually, there are numerous communities within any specific formation, and the identification of regional community types which might be equated with formation boundaries is often quite uncertain making any further correlation useless.

The Importance of Formation

Because of the emphasis on plant composition, the importance of formation becomes mainly its role as the habitat for animals. But animals include economic man so the formation has much to tell about use potential just as have all the other aspects of the geography of life. When plants and animals are functionally combined, as with formation and inhabitants, the concept of a system, the subject of the next chapter is approached.

Habitat

Each of the seven general formations and any subdivision thereof or other type constitutes a habitat for animal life. Some animals, because of their feeding habits, may be able to subsist in several formation regions. Others are so specialized as to require a certain plant group for subsistence. Certain general types, however, can be identified by their characteristic formation. There are, to be sure, the water habitats such as rivers, lakes, and the various ocean environments not discussed in this chapter. Most of these are special types, but, with more study, the major marine habitats might effectively be incorporated under the formation approach.

Rainforest

With the food concentrated high in the canopy, most rainforest animals are either climbers or fliers. The vegetarians range from insects and birds of great variety to bats, monkeys, and sloths. Predators are represented by such forms as snakes, leopards, and other species of insects and birds. Not many large animals are to be found. On the ground there are only a few scavengers such as tapirs and cassowaries.

Seasonal Forest

Although resembling the rainforest, the seasonal forest offers more food at ground level and has few large climbing animals. Squirrels and rats are characteristic together with such forms as bears and raccoons. Migration and hibernation are common adaptations along with various food storage strategies. Most rainforest animal types are present. With disturbance, all manner of woodland animals penetrate the forest zones.

Woodland and Savanna

A mixture of trees and ground cover produces a major basic source of food called *browse,* sometimes indicated by the term *edge.* Cattle, deer, and similar animals harvest this resource. The elephant and the giraffe are some more spectacular browsers. One might also associate the anteaters with such mixed vegetation. Small tree animals such as squirrels, birds, and insects continue with the presence of trees. Predators begin to reflect the open country opportunities. The woodland probably includes the greatest range of habitat choices of any formation.

Grassland

The basic food source of a grassland is grazing. The plant eaters here characteristically run in great herds, as much for mutual protection as because of the vast monotony of the environment. One thinks of sheep, horses, antelope, and bison. Even the predators tend to work in groups or packs. Consider dogs and hyenas or even lions. Grassland animals are noted runners and a predator, the cheetah, is the fastest of all. Other outstanding runners are ostriches and jackrabbits.

Desert

More than anything, the desert habitat has reduced resources. Many animals of the grassland and woodland types also survive in the desert. Accumulating food and doing without water are desirable abilities for desert animals, but because of their mobility, animals are less restricted than plants. The packrat is a desert dweller and so is the camel. But few desert animals differ significantly from those of other habitats.

It must not be imagined that the brief examples given here represent all animal forms or all habitat possibilities. Water animals include browsers and predators, both large and small. A great many sea animal groups are poorly represented if at all on land, but the basic feeding habits and strategies are largely similar.

Early Man

At some prehistoric time man changed his subsistence through the use of tools. For early man this did not make his relationship to habitat much different from other animals but it enabled him, by use of appropriate tools, to penetrate any food-containing habitat. Some of the races of middle paleolithic times still survive today and it is of interest to compare their original distributions with environmental patterns (Map 6-3). Three groups can be identified.

Negrito

Only fleeting traces of the early negrito, sometimes called pygmy, are available but their distribution can be rather reliably reconstructed. Two primary negrito regions exist, one in the Congo basin and the other in the East Indies (southern China to New Guinea). Negritos have also been detected on the west coast of India, the east coast of Australia, and in Tasmania. All of these areas support rainforest and a particular kind of human food potential. Essential protein is concentrated in water bodies and in the forest canopy. Negrito economy is heavily oriented to fishing with hooks, harpoons, nets, and poison. The canopy is harvested with darts and poison arrows. The negritos early specialized in rainforest food gathering, apparently also simultaneously differentiating racially. Their place of origin may have been Africa, but they subsequently were able to populate virtually all of the Old-World rainforest habitats. Their culture did not allow them to subsist in more open country.

Australoid

A variety of similar early peoples sometimes grouped together under the name Australoid once occupied a great deal of the world. Their two primary concentrations are in India and Australia, where some still survive, but they also have occupied nearby areas. Some experts think that they once extended into Africa while the earliest men in America belong essentially to this group. Perhaps the Ainu represent a diluted remnant in east Asia. Earlier man had mastered fire, and by middle paleolithic times most seasonal forests were opened up to resemble woodlands in general. All of the range of the Australoid lay in open country where he hunted with spear and throwing stick or with snares.

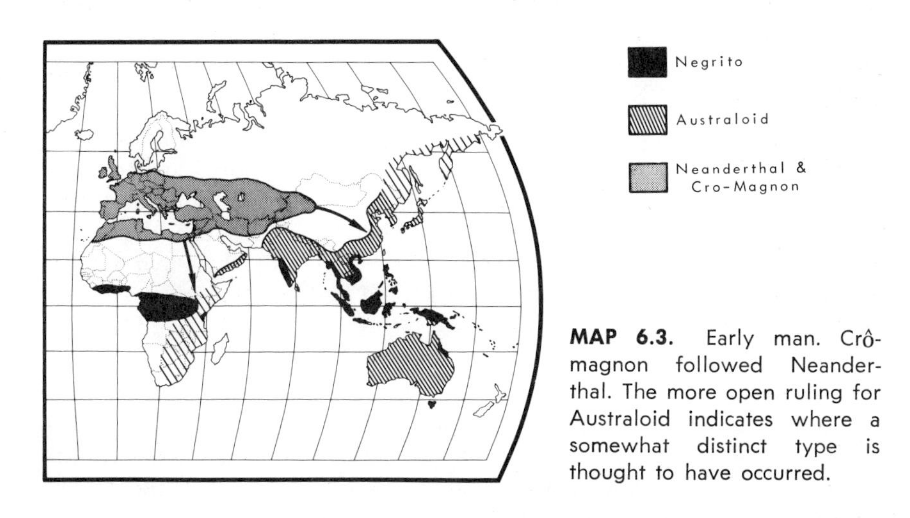

MAP 6.3. Early man. Crô-magnon followed Neanderthal. The more open ruling for Australoid indicates where a somewhat distinct type is thought to have occurred.

The round-up, sometimes using fire, was a very productive source of food even if it was wasteful. The Australoid early adapted to his kind of environment and spread far and wide, probably throughout the world wherever open country was to be found. The rainforest would yield little food to his methods and so he and the negrito easily maintained their racial differences.

Neanderthal and Cro-magnon

It was in colder parts of the world that the now extinct Neanderthal held sway. His hunting grounds lay next to the great glacier from the Atlantic to central Asia. This habitat, except for the cold, resembled that of the Australoid, and Neanderthal conquered it partly by using clothes. Some 30,000 years ago Cro-Magnon man replaced Neanderthal throughout his range and then expanded to additional territories. Cro-Magnon continues today as an important element of the Caucasian peoples.

The early races of modern man occupied all of the world's land area during the last glacial period with three cultures and three similar physical types. Only later was cultural mobility sufficiently advanced for members of a single population to be able to occupy any part of the world, that is, for racial distribution to be freed of habitat limitations. It is probable that habitat isolation was still mainly responsible for the physical differentiation of man in the middle paleolithic. Later men, Negroes, Caucasoids, and Mongoloids in particular, with more advanced cultures, each rapidly overran territories of vastly contrasting environment.

Economy

In spite of man's adaptability today, the different conditions that are to be met from place to place on the surface of the earth cannot be overlooked. From the point of view of formation, the potentials and limitations of different areas can be understood. Because there are interactions between vegetation and other aspects of the environment such as climate and soil, the importance of a formation area can be expressed partly in these other terms.

Rainforest

In spite of the lack of unfavorable seasons for plant growth, rainforest areas present serious difficulties. In the first place, the lack of unfavorable seasons applies equally to pests, insects, fungus, and others as it does to man or his domesticates. Second, rainforest soils are inherently leached sterile and are virtually unusable for crop production in the usual sense. In fact, the whole complex of cropping developed in the civilized world is unsuitable for the rainforest environment. That there are effective ways of using these areas is largely academic and if they are to be used to advantage new techniques will have to be developed.

Seasonal Forest

Man has been most successful in seasonal forest areas. Production with his domesticates, plants, and animals is highly suited to this environment. The cycle of production and storage lends itself well to the exploitive needs of man. What this amounts to is that man has thoroughly adjusted to the problems of seasonal forest regions, problems like survival during the unfavorable season and managing the poor soil.

Woodland

In some ways woodlands (and similarly *brush*) are less of a problem than seasonal forests, in others they are more, but for man their use has been similar. The soil is definitely better in the drier woodlands (but not in the cold woodlands); the briefer favorable season tends to limit production possibilities. Because woodlands are mostly in broken terrain and are rather scattered, their usefulness is reduced, although pockets of alluvial soil can result in local highly productive areas as around the Mediterranean. Irrigation can be used to overcome unreliable and deficient moisture supplies.

Desert

The extreme conditions of deserts can make them unusable under most available technologies. Cold deserts particularly are intractable but warmer areas usually yield handsomely to irrigation where it is available. Deserts tend to offer freedom from humidity and storms so that, for some advanced activities, they may become desirable.

Grassland and Savanna

Being disturbed formations, grasslands and savanna partake of many of the properties of their corresponding undisturbed types. They offer several aspects in their own right, however. These herbaceous covers are natural rangelands for livestock herds and, since the advent of the neolithic, man has brought these animals under control. Grassland soils tend to be rather fertile making these areas desirable for agriculture in spite of some moisture deficiency. The characteristically level terrain also lends itself to large operations and economies of scale.

There are many other factors to economy than those outlined above, but these alone taken together with current objectives and technical abilities explain a great deal of man's productivity. Measured in population, one can observe that rainforest and desert are essentially empty and seasonal forest is, with few exceptions, well populated. Woodlands and the disturbed formations (largely alternating with woodlands and seasonal forests) have intermediate to dense populations. Technological advances may emancipate man increasingly from environment, but environment still matters and the basis upon which advances must be built has developed in harmony with the environment.

Chapter 7

Ecosystems

The complex of life together with its environment on the surface of the earth constitutes the ecosphere, a dynamic interacting network considered as a whole more than by the parts. The ecosphere can be subdivided into distinct units, each of which is called an ecosystem, of which there may be many varieties. The concern here is for the geographic aspects of these systems as they relate to plants and animals, always remembering that there is much more encompassed in the study of ecosystems and that not all aspects of the biosphere are systematically interrelated. Geographically, the study can be divided into investigations of the regional expression of such systems, their spatial reality, and their conceptual value.

Geographic Expression

Ecosystems manifestly exist in space and all aspects of them undoubtedly have a distinct geographic expression. The role of plants and animals as constitutents of the systems can be examined with reference to the concepts advanced in previous chapters; this should then lead to an understanding of the areal differentiation of the whole ecosphere. There follows the possibility of differentiating regional kinds or varieties of ecosystems of which the more important categories will be identified.

biological composition

In the previous chapters, several kinds of life aggregations were developed and, inasmuch as each displays systematic properties, they must of necessity become elements of ecosystems. The interacting complexes which have been named are the community or biocenose, the formation, and the biosystem or biome. Something of a structural hierarchy is involved here leading finally to the whole ecosystem.

Community (Biocenose)

To the degree that distinct communities can be distinguished, they form the ultimate division of the ecosystem. Each community is distinct and each community has a corresponding environment, as for example a swamp or a mesic upland. By grouping communities in terms of similar habitats, as is often done, the effect is to abandon the life aspect of the ecosystem and to proceed through the environment. The grouping of communities in terms of taxonomic relationships produces the biota whose framework is biologic evolution, outside of the kinds of relationships operating within ecosystems. The only way to treat communities as parts of an ecosystem, therefore, is as unique elements or phases which, it has been argued, may not correlate well with the larger aggregates.

Formation

Essentially the formation is the basis for ecosystem differentiation because the emphasis of the formation is on plants, the photosynthesizers or food producers upon which all other life is dependent. Plants also, whether they are rooted on land, floating at sea, or growing on the ocean bottom, comprise the greatest bulk of life in an area and their form characterizes all of life as it interacts with the environment. Formation differences, therefore, represent functional divisions of the ecosphere and each kind of formation is systematically distinct.

Biosystem (Biome)

The addition of animals to the formation comprises the biosystem. Although animals are essentially parasites on plants, they function as regulators and sometimes partake of beneficial symbiotic relationships with plants. The basic food of animals is plants and, without animals, the biotic content of each area would be quite different. Not only selective consumption but also pollenation and seed preparation by animals makes a great deal of difference how much of what plants will grow. Whereas drainage and other conditions might not be reflected in plant formation differences, they are important to the nature of animal life so that, in this respect, a single kind of plant formation can be considered an aggregate of several kinds of biosystems.

Ecosystems

By adding the environment to the biosystem, one finally has the ecosystem. The biosystem reflects the environment, of course, so the distinction may be somewhat technical. In any case, each part of the ecosystem has its effect on all other parts and, taken separately, a series of studies can be made. Taken as a whole, the interaction, such as nutrient cycles, can be comprehended. Nor can any of the parts be completely understood without reference to the whole system in which it lies. Any areal variations of the elements will also be reflected in the whole. With this in mind, attention can now be directed toward the identification of the major varieties of natural ecosystems.

varieties

Division of the ecosphere involves differences from place to place, each distinct areal segregate being an individual ecosystem and, considered geographically, also a region. Vertical distinctions (synusia) and other kinds of separations can also be made and have been recognized by ecologists. Regional ecosystems may be large or small, reflecting the patterns already discussed that are displayed by the constituent parts of the ecosystem, and in similar ways can be grouped into types. Two general categories of regional ecosystems are suggested, a local or special group and a zonal or general sort. It is possible that some areas, particularly as a result of disturbance, may be occupied by life not interacting in a systematic way and, together with lifeless places, such areas would not be considered to support an ecosystem.

Local

Significant contrasts within short distances of the ecological conditions are associated with slope, exposure, or underlying material. Soil or edaphic factors and terrain-form may change abruptly with a corresponding change in the community structure and possibly also the biosystem. Such discontinuities tend to produce distinct local ecosystems that can be regarded as pockets or islands, neat packages readily amenable to study.

One set of local ecosystem types is related to drainage. The depth and duration of excess water together with aeration and substratum makes for a spectrum of types from swamp and bog to several kinds of lake. Where trees are absent, one speaks of marsh. The poorly drained area is not at all everywhere uniform, there often being zones in which a series of types from the margin to the center can be seen. All sorts of edge effects are also possible as for animals which live in a marsh but forage in the nearby forest or even the reverse of this. Excess drainage can be just as important as poor drainage but tends to parallel drier climatic responses. Some ridge communities can be related to poor water retention.

Another set of local ecosystems are coastal. Very narrow and intermittent strips lie next to one another between the sea and dry land. A rich array of plants and animals, such as coconuts, sea grapes, crabs, plover, seaweed, and many others, each takes its special position along the coast. As the coastline varies in structure, so the ecosystems change. Sheltered estuaries and lagoons in winter climates have salt marsh but several tree genera have conquered this difficult environment in the tropics to produce mangrove. Tropical coasts also offer the phenomenon called a reef, a most intricate ecosystem. The area of coastal strips, when totalled amounts to an impressive portion of the world's surface.

Local soils, without considering drainage, offer many strong contrasts based on differing parent material. Pockets of aggradation may bring, a rich, dense medium next to entirely different substances. Structure may affect internal drainage and mimic climatic differences, but chemical content may work remarkable changes. In many widely separated localities there exist local accumulations of dense powdery soil poor in mineral content, and here survives an impoverished and often stunted community quite different from that of surrounding areas. Outcrops can produce the same effect. Not only is bedrock reflected in the overlying soil structure but also in the mineral content. Soils over serpentine exposures often also support a curious pygmy forest. Volcanic deposits tend to weather into rich soils which, if they lie beside a granite area, support a much more luxuriant community than that possible with the poor granite-derived soils.

Running water produces a series of environments related to velocity and sediment content. Essentially, there is a series from head to mouth which divides in terms of sediment. Clear mountain streams become murky organic-filled rivers or sluggish mud soups that end in deltas or under the sea where finally the load comes to rest. The mixing of fresh and salt water has its effect on local ecosystems and it must be realized that vast amounts of organic foods are disgorged along with sediments to support a rich fauna even where plants may not maintain a hold.

Other local effects can be imagined. Exposure of one sort or another can have profound results. Elfin forests owe their development to a particular set of exposure factors in a very humid climate along a ridge. Other ridge communities may be responding strongly to wind as in the case of certain dense ridge thickets. Exposure can merely result in strong local climatic contrasts as between the exposed and sheltered sides of a ridge with respect to sunshine, the one side warm and dry, the other cool and moist.

Each local ecosystem has its peculiar community by which it differs from surrounding parts of the ecosphere. In terms of formation such places were called special types, where distinctions could be recognized, but these local ecosystems often do not form contrasting or even distinctive kinds of formation. It is the local isolation of species, therefore, that most characterizes this kind of ecosystem. Comparing the many individual examples there may be numerous occurrences of the same or essentially similar communities just as there may be many islands in an archipelago.

Zonal

Where local variations do not intrude, the ecosphere may display general uniformity over impressive distances with the world at best being divided into broad ecological zones. For the most part, these zones are responding to climatic variations and are characterized by the plant formations. Within each biotic realm there is a specific community for each zonal division and structurally similar zones in different realms can be grouped into kinds of zonal ecosystems. A major distinction can be made between continental and marine types.

The continental environments have been well characterized by plant formations. By combining the resident animals with the corresponding plant formation, their habitat, one has a biosystem or biome as it has been called. Thus there are, among others, a rainforest biosystem and a seasonal forest biosystem. It must be understood in these biosystems, as in all others, that considerable internal variation is present over the large distances that are involved even though these variations are gradual and relatively mild. Naturally, such large ecosystems have been difficult to comprehend for study.

Marine environments have been less researched than land ones but several clear divisions can be made. The floating or pelagic environment easily contrasts with the bottom or benthic habitats even though its life is derived from benthic ancestors. Aside from special local ecosystems, life on the bottom is made up of littoral communities where there is light and abyssal scavengers in the deeps. Terms such as underwater meadow and forests of seaweed have been applied to littoral formations, and it is likely that much the same reasoning which has been used with land formations would obtain here. There must be sea-bottom woodlands and the abyssal is, in fact, a form of desert.

There even seems to be a zonal aspect to the local ecosystems. Within a specific zonal system, many of the local communities tend to conform to particular types which differ from the types within other zones. Swamps responding to a particular drainage situation within a particular forest zone are generally all alike but equivalent communities within the woodland tend to be different. Perhaps there are such zones for certain local ecosystems that do not coincide with the regional systems but have their own distribution. There is ample room for further study of the geography of ecosystems.

Geographic Reality

The ecosystem idea is a concept the reality of which can be questioned. To demonstrate, at least partially, the validity of the concept is also to suggest why such distinctions occur. A system is a set of elements that have definite relationships and interactions, not just a random juxtaposition. Evidence to confirm the existence of a system can, from the point of view of geography, involve the coincidence of the elements in space and the presence of recognizable boundaries. These properties have not always been demonstrated, although it should be immediately realized that there is probably no such thing as a completely self-contained or endogenous system, certainly not on the surface of the earth. The relationships can only be relative but they must be more than random.

Areal Integrity

Intuitively or as a result of informal observation it can be said that distributional relationships exist between plant and animal species on the one hand and their environment, soil, climate, and surface features, on the other. The nature of this relationship is not always clear, and conflicting interpretations, not to mention arguments, have appeared on the subject. The biota-enviroment relation can be considered first and a review of the community nature will also be pertinent.

Soil owes its very origin in part to a contribution by life. Dead and decaying organic material is a constituent part of any soil along with a living element, roots, bacteria, fungus, and even burrowing animals. But soil is only in part organic. It has, as well, properties derived from its parent material which are further modified by surface configuration. To be sure, these things have their effect on the life forms that are present, but it is too much to equate soil with the biosystem. There are aspects of soil, content and structure, which do not coincide with the patterns of resident life even though soil makes an important contribution to the corresponding ecosystem.

The impact of climate on life is nearly everywhere obvious; the impact of life on climate is not so clear. Plants can only be active with temperatures above a critical level and with an adequate moisture supply. Climatic stress was invoked in discussing formation differences. On the other hand, there are many aspects of climate that have little or no direct effect on plants, such as precipitation cause, storm structure, or pressure system.

It can be said that climate and soil or, to a lesser extent, other factors such as surface configuration are involved with the biosphere but are not united with it. If certain environmental parameters can be identified which directly affect life systems or are produced by them, these can be considered as a part of the ecosystem and geographic correlations should be expected. The environment as a whole cannot, however, be included in organic "natural regions" or even in ecosystems as has sometimes been done in the past.

The further question concerning regional ecosystems is how well the associated species in any area can be linked into a system. All too often no significant relationship has been adduced, the community being but a casual overlap of distributions. Involved here is the reality of the community concept which was discussed in Chapter 4. Some schools of thought would attach virtual organismic status to natural communities while others completely deny their existence. The best that can be said is that there is doubt about the internal areal integrity of many reputed ecosystems.

Boundaries

Another approach to defining an ecosystem is to discover boundaries or discontinuities which segregate one system from another. Such boundaries need not be absolute to confirm that a distinct system exists. Happily, all sorts of limits have been identified and these will help in verifying the reality of ecosystems. Boundaries can be conceived either as having been externally imposed or internally developed.

Many naturally occurring discontinuities in the biosphere have resulted from external causes. Impermanent boundaries produced by disturbance or by other factors are probably of no systematic significance unless they coincide with limits produced in other ways. Soil and terrain discontinuities, however, can be quite sharp as well as permanent. Some of these are random and are not amenable to classification but others are quite specific. The biosphere is inevitably affected by environmental contrasts and will likely be separated thereby into distinct assemblages. It might be argued that such divisions are forced and the resulting groupings still are not necessarily systematically interrelated. But here it is a question of a genetic condition, permanent and specific, so that through time particular interactions would be expected to develop where they did not already exist. Stable kinds of environmentally imposed isolations therefore correspond to systematic divisions of the biosphere. These are, of course, the local ecosystems which have been described above.

Boundaries in the biosphere not related to environmental discontinuities or produced by disturbance have been described. There are not many examples of these, but they are quite important. Primarily, the reference here is to formation boundaries which tend to divide rather distinct life structures on the surface of the earth. These are the zonal ecosystems and their reality rests in large measure on the existence of formation boundaries because not enough has been done geographically to characterize such areas by content. That is, in dealing with the major biosphere zones, too little attention has been given to the internal place to place variation. Floristic and faunistic divisions have also been mentioned. Where these are sharp is where disturbance has failed to destroy the related formation boundaries. That is, biota tends to divide along formation boundaries and probably is not responsible for the differentiation of any distinct ecosystems.

Geographic Value

The geographer as well as anyone else who grapples with aspects of life on the surface of the earth finds that the systems approach helps in the comprehension of complex reality. The analytical approach, holding other factors constant while only one phenomenon is studied, has its value and has been widely pursued, but the real world exists as a whole with dynamic interaction. The ecosystem concept offers a technique with geographic relevance for solving problems relating to the biosphere with more attention to its inherent complexities than might be possible with a straight analytical approach. Regional dynamics can thus be given a thorough inspection.

Methods of Research

The ecosystem is a paradigm for research which geographers can use. This can be regarded in two ways: the contribution that the paradigm has for geographic problems and the contribution that geography has for ecosystem problems.

With the ecosystem, an avenue is constructed for the introduction of all of the thinking of systems theory to the geography of the biosphere. Quantifiable relationships are suggested and all that that means in research procedure. Linkages and chains, inputs and outputs, and feedback all have their geographic aspects. Not that questions of a similar import have not been posed before, but with a paradigm the investigation promises to be more thorough. And with this approach to the natural ecosystem, the next step is transferral of the whole design to the cultural ecosystem. Here is a research frontier.

Geography has much to offer through the ecosystem to studies in other disciplines. Mention was made of the random juxtaposition of populations or, for that matter, of any two phenomena on the surface of the earth. Proximity does not guarantee interaction, even indirect, and yet it is easy for a non-geographer to suppose that neighbors are in communication. A demonstration of the distributional relationships of phenomena is one way of getting at their interactions, if any. The timing and sequence of interactions also depend on relative position, another situation in which a geographic awareness is required. Clearly, the investigation of ecosystems has much to profit from a geographic approach, whatever it may be called, and much to lose without it.

Regional Dynamics

Regions are a fundamental geographic generalization and regions are dynamic objects. Considered in this way it might even be said that regions are ecosystems; certainly many are. The dynamic properties of regions can be handled by systems methods. Of the many dynamic aspects of regions, two overriding considerations are seen today, production and pollution.

Production is another way of talking about input and output. In the natural ecosystem, there are various possible inputs and outputs such as rain, sunshine, dust, and excess organic matter. It is simple to extract from an ecosystem on an exploitive basis. It is more difficult to plan a sustained yield such as livestock products or lumber harvest. Excessive demands on an ecosystem will produce destructive changes which must be avoided where a permanent economy is desired. Development of an area also depends on production. In nature, dynamic succession is brought about through the accumulation of outputs which achieve environmental threshholds. Man can manipulate ecosystem growth for his own purposes in a similar way and it is so much more rational if he understands the systematic relationships of the relevant factors. Neither is production nor are ecosystems uniform from place to place so the role of geography in production planning is clear.

Pollution is a sort of feedback. The manipulation of an ecosystem will produce a chain of reactions, some of which may be undesirable. In the world today, environmental manipulation has pyramided without commensurate attention to the possible reactions. But the reactions inevitably occur. Nature is so complex that there is no hope of anticipating all such reactions, but there is much that can be done to reduce the number of unpleasant surprises. Because there is a geography to ecosystems, geographers must take part in the pressing job of pollution control. Areas of runoff and erosion yield volumes of flood and sedimentation. Remaining range area or distribution tells of survival or extinction. Directions of movement, build-up, local over-use, all of these things are very geographic and geographers must be called upon to help give the necessary information about them.

Bibliography

General

CARLQUIST, SHERMAN, *Island Life*, Garden City, N. Y.: Natural History Press, 1965.

EDWARDS, K. C., "Importance of Biogeography," *Geography*, Vol. 49, pp. 85-97, 1964.

JUST, THEODOR, "Plant and Animal Communities," *American Midland Naturalist*, Vol. 21, No. 1, 1939.

NEILL, WILFRED T., *The Geography of Life*, New York: Columbia University Press, 1969.

NEWBIGIN, MARION I., *Plant and Animal Geography*, London: Methuen & Co., Ltd., 1936.

SIMPSON, GEORGE G., *The Geography of Evolution*, New York: Capricorn Books, 1965.

VORONOV, A. G., "Biogeography Today and Tomorrow," *Soviet Geography: Review and Translation*, Vol. 9, pp. 367-377, 1968.

WILHEIM, E. J., "The Role of Biogeography in Education," *Journal of Geography*, Vol. 67, pp. 526-529, 1968.

Plant Geography

BOYKO, H., "On the Role of Plants as Quantitative Climate Indicators and the Geo-Ecological Law of Distribution," *Journal of Ecology*, Vol. 35, pp. 138-157, 1947.

BRAUN, E. LUCY, *Deciduous Forests of Eastern North America*, Philadelphia: The Blakiston Co., 1950.

BRAUN-BLANQUET, J., *Plant Sociology*, New York: McGraw-Hill., Inc., 1932.

CAIN, STANLEY A., *Foundations of Plant Geography*, New York: Harper & Brothers, 1944.

CARTER, GEORGE F., "The Role of Plants in Geography," *Geographical Review*, Vol. 36, pp. 121-131, 1946.

CHIKISHEV, A. G., ed., *Plant Indicators of Soils, Rocks, and Subsurface Waters*, New York: Consultants Bureau, 1965.

COLE, MONICA, "Distribution and Origin of Savanna Vegetation in Brazil," *Geographical Journal*, Vol. 126, pp. 168-179, 1960.

CROIZAT, L., *Manual of Phytogeography,* The Hague: Dr. W. Junk, 1952.

DANSEREAU, PIERRE, *Biogeography,* New York: The Ronald Press Company, 1957.

EYRE, S. R., *Vegetation and Soils,* London: Edward Arnold, 1963.

GLEASON, HENRY A., and CRONQUIST, ARTHUR, *The Natural Geography of Plants,* New York: Columbia University Press, 1964.

GOOD, RONALD, *The Geography of the Flowering Plants,* 2nd ed., London: Longmans, Green and Co., Ltd., 1953.

HILLS, THEODORE L., "Savannas: a Review of a Major Research Problem in Tropical Geography," *Canadian Geography,* Vol. 9, pp. 216-228, 1965.

KUCHLER, A. W., "Plant Geography," Ch. 19, JAMES, P. E., and JONES, C. F., *American Geography, Inventory and Prospect,* Syracuse: the University Press, 1954.

———, *Vegetation Mapping,* New York: The Ronald Press Company, 1967.

DE LAUBENFELS, DAVID J., "The Variation of Vegetation from Place to Place," *Professional Geography,* Vol. 20, pp. 107-111, 1968.

MCINTOSH, ROBERT P., "The Continuum Concept of Vegetation," *Botanical Review,* Vol. 33, pp. 130-187, 1967.

POLUNIN, NICHOLAS, *Introduction to Plant Geography,* New York: McGraw-Hill, Inc., 1960.

RICHARDS, P. W., *The Tropical Rainforest,* Cambridge: The University Press, 1952.

RILEY, DENIS, and ANTHONY YOUNG, *World Vegetation,* New York: Cambridge University Press, 1966.

WHITTAKER, ROBERT H., "Classification of Natural Communities," *Botanical Review,* Vol. 28, No. 1, January-March, 1962.

———, "Dominance and Diversity in Land Plant Communities," *Science,* Vol. 147, pp. 250-260, 1965.

Animal Geography

ALLEE, W. C., and SCHMIDT, KARL P., *Ecological Animal Geography,* 2nd ed., New York: John Wiley & Sons, Inc., 1951.

DARLINGTON, PHILIP J., *Zoogeography,* New York: John Wiley & Sons, Inc., 1957.

DAVIES, J. L., "Aim and Method in Zoogeography," *Geographical Review,* Vol. 51, pp. 412-417, 1961.

DICE, L. R., *The Biotic Provinces of North America,* Ann Arbor: The University of Michigan Press, 1943.

GEORGE, WILMA, *Animal Geography,* London: William Heinemann, Ltd., 1962.

KENDEIGH, S. CHARLES, *Animal Ecology,* Englewood Cliffs, N. J.: Prentice-Hall, Inc., 1961.

STUART, L. C., "Animal Geography," Ch. 20, JAMES, P. E., and C. F. JONES, *American Geography, Inventory and Prospect,* Syracuse: Syracuse University Press, 1954.

Ecosystems

DARLING, F. FRASER, and MILTON, JOHN P., *Future Environments of North America,* Garden City, N. Y.: Natural History Press, 1966.

FOSBERG, F. R., ed., *Man's Place in the Island Ecosystem: a Symposium,* Honolulu: Bishop Museum Press, 1963.

MORGAN, W. B., and MOSS, R. P., "Geography and Ecology: the Concept of the Community and its Relationship to Environment," *Annals of the Association of American Geographers,* Vol. 55, pp. 339-350, 1965.

OVINGTON, J. D., "Quantitative Ecology and the Woodland Ecosystem Concept," *Advances in Ecological Research,* Vol. 1, pp. 103-192, 1963.

SHELFORD, VICTOR E., *The Ecology of North America,* Urbana: University of Illinois Press, 1963.

STODDART, D. R., "Geography and the Ecological Approach: the Ecosystem as a Geographic Principle and Method," *Geography,* Vol. 50, pp. 242-251, 1965.

———, "Organism and Ecosystem as Geographical Models," Ch. 13 in CHORLEY, R. J., and HAGGETT, PETER, *Models in Geography,* London: Methuen & Co., Ltd., 1967.

TOSI, JOSEPH A., "Climatic Control of Terrestrial Ecosystems: A Report on the Holdridge Model," *Economic Geography,* Vol. 40, pp. 173-181, 1964.

Index